Boundaries of Evolution

An Analysis Showing the Limitations of Evolutionary Theory

What Would Darwin Think
Now About DNA, the Big Bang,
and Finite Time?

Theodore R. Johnstone, M.D.

Order this book online at www.trafford.com
or email orders@trafford.com

Most Trafford titles are also available at major online book retailers.

Cover Design or Artwork by Mike Troup
Designed by Jessica Garcia
Edited by Judy Coulston Ph.D.
Illustrated by Jessica Garcia

Print information available on the last page.

ISBN: 978-1-4907-4566-4 (sc)
ISBN: 978-1-4907-4567-1 (e)

Library of Congress Control Number: 2014915636

Trafford rev. 07/06/2017

www.trafford.com

North America & international
toll-free: 1 888 232 4444 (USA & Canada)
fax: 812 355 4082

Acknowledgements

Without the patience of my lovely wife Kitsy, to whom I have been married for more than a half century, the self-imposed endeavor of writing this book would have met an early demise! Kitsy has put up with much through the years, especially after we obtained a home computer equipped with Microsoft Windows XP and Dragon Naturally Speaking with voice recognition. This scientific electronic marvel has allowed me to dictate text directly into my computer. Before this, lacking typing skills, I had to write and rewrite many pages in longhand. I then recorded these handwritten pages on tape and sent them for transcription to my secretary Judy Caudill. She too was very patient with me as I stumbled along. Several years before that, Alice Munoz did some typing for me. Before I got a little computer savvy, and was able to take advantage of Microsoft Word Perfect and Spell Check, two other secretaries, Marilyn Stranland and Barbara Sallee (each now having succumbed to an untimely death), helped me multiple times with spelling and grammar. Marilyn's daughter Gail Dummer also did some typing. My roommate from college with a Ph.D. in mathematics, Lawrence Hanson -- now retired from teaching college mathematics, has been of great assistance. The former chairman of the Department of Mathematics at California State University Fresno, Ronald Wagoner Ph.D., also helped me with some mathematical calculations. Judy Coulston, with her Ph.D. in nutrition was extremely helpful, not only in her area of expertise, but also in editing the text itself. Albert Brown, M.D., a pathologist, Brian Bull, M. D., a pathologist, Jared Verner, Ph.D., a biologist, Jerry Guibor, a retired journalist, Walter a student at an Ivy league school and Mr. Puma, a chemist, all have read my manuscript and given many helpful suggestions. Judy Caudill and Jessica Garcia, our fourth daughter, each have contributed many of the diagrams found in most chapters. In addition, my friend Daniel DeSantis, has given me other suggestions. Without the encouragement and help of these wonderful people, this book would never have come to fruition. They deserve a standing ovation.

T.R.J.

INTRODUCTION

It is easy to believe in science. The scientific method of investigation using research and experimentation or observations has produced many wonderful discoveries that range from penicillin to men walking on the moon. When scientists speak, we listen! But are they always right?

I am a country doctor, having practiced in the small town of Madera California since May of 1968. I graduated from medical school in 1959, during which time I had only one lecture on the structure of DNA. This lecture occurred while I was a freshman medical student in the academic year1955-56. The genetic code contained in the DNA was then about a decade or more away from being cracked. In the more than five decades that I have had the privilege of practicing medicine, many scientific medical advances have been made. Many of these advances I have had the privilege of incorporating into my own practice for the benefit of my patients.

In 1955, I graduated from college with a B.A. degree in chemistry and a minor in physics. Later in medical school, when the complex structure of the DNA molecule was first described to us, the thought crossed my mind then, and has persisted to this day that it would be very unlikely for DNA to self construct in the primordial soup with no controlling factors. This notion was based upon my frustrating experiences of synthesizing organic molecules in the lab under controlled conditions. Therefore, doubts arose in my mind as to what science had to say about the origin of life.

During my on going tenure in the practice of medicine, I have seen many so-called scientific principles that I learned in medical school discarded by new research contradicting my former instruction. One of the most outstanding was what I was taught about the etiology and treatment of duodenal ulcer. Back in med school we were taught that duodenal ulcers were precipitated by worry and stress. In turn, under the influence of the autonomic nervous system, the stomach secreted excess hydrochloric acid, eventually causing the mucosa of the duodenum to break down, resulting in an ulcer. Treatment centered on antacids, medication to decrease the amount of acid that the stomach secreted, and medication to decrease the amount of worry. This approach did help many patients but a considerable number ended up in surgery with various surgical approaches to the problem. Sometimes surgery worked and sometimes it didn't!

Then way down under, out in Perth Australia, two men, one a pathologist and the other a medical student by the names of J. Robin Warren and Barry J. Marshall, suspected that duodenal ulcers might be caused by a bacterium now known as Helicobacter pylori. To prove their hypothesis they devised a new treatment using antibiotics to eradicate this noxious bacterium. To their amazement, when H. pylori was eradicated from the stomachs of infected patients, the ulcers healed. At first the scientific community ridiculed their published findings. However, to add further credence to their discovery, Marshall actually drank a cocktail of H. pylori, became ill but later recovered with their treatment! Finally, somebody paid attention to their breakthrough discovery. After being invited to the U.S. to explain and demonstrate their findings and treatment, their work was accepted by the scientific community, for which they received a Nobel prize in 2005. Even after their work was recognized, it took some time for this new information to filter down to those of us on the front lines of medicine. I can still remember the last patient I sent to a general surgeon for ulcer surgery. That was before either one of us had heard about the actual cause and new treatment for ulcers. I could cite multiple other examples from my medical practice; however, this one is cited here only to show that science is not always right. I had to unlearn what had been taught to me as scientific fact.

The October 1994 issue of *Scientific American*, a journal to which I have subscribed for many years, devoted its entire issue to the theory of evolution. Remembering back to the one and only lecture on DNA that I had received in medical school and how I didn't see then how this complicated molecule could self-construct, this issue of *Scientific American* started me on a scientific journey that eventually culminated in the writing of this book. For years before and after 1994, I had been working 10 to 12 hours per day plus being on call 24 / 7, so I did not have much spare time. The little intermittent spare time that I did have, was spent mostly at night in countless hours reading and rereading multiple scientific books and scientific journals educating myself about the DNA molecule. This study has brought me up to the twenty-first century with regard to a better understanding of such things as viruses, prokaryotes, single-celled eukaryotes, and multi-celled eukaryotes. My study has also helped me to better understand some of the basic science surrounding newer medications, some of which are stereoisomers with left-handed configurations. However, after becoming semi-retired, I took advantage of the increased time available and set about to finish the manuscript.

The media, as well as high school and college textbooks, along with scientific periodicals present evolutionary theory as though it rivals Einstein's theory of relativity with undisputed proofs. However, new discoveries and recent research are casting considerable doubt on the veracity of evolutionary theory regarding its explanation of biota that lives on this planet. This book is an attempt to review these boundaries objectively, allowing the serious reader to come to a reasonable conclusion. The main agenda is to get all interested persons to take an objective second look.

The first two chapters give a historical perspective. Chapters three through eight are devoted to teaching the minimum basic science needed to understand the scientific limitations constraining evolutionary theory that follow. The summaries that follow most chapters are concise and are designed to assist all who might need some help grasping the data presented.

If you had lived at the time of Galileo, would you have been one of those who refused to look through his simple telescope to see what he saw, or would you have refused to believe what you saw, even if you had looked, or would you have looked and believed? Like the educated elite of Galileo's time, some modern counterparts will refuse to view the evidence.

I hope that everyone who reads this book will be challenged by the evidence. To paraphrase Gerald Schroeder, who received his Ph.D. in physics from MIT: Cherished axioms die hard even in the presence of overwhelming, contradictory evidence.

T.R.J.

Table of Contents

Chapter 1

The Birth Pains of Science

Maybe it started only by a casual glance, a chance encounter, or even a formal introduction, but the chemistry was there. Perhaps the romance took only days or weeks, or possibly it took years. The courting may have occurred unconsciously at first, or the progress was slow, but as feelings were nurtured, the embers that seemed at times to almost go out suddenly burst into flames. Two rings, two vows and the two became one, resulting in time in birth pains and delivery of a new person.

The romance between two people is not unlike the progress of science. An idea in someone's head brought on by a casual glance at something, or a chance encounter as when an apple falls from a tree, or a formal introduction in a classroom setting starts the idea growing and over time produces a scientific concept and birth of a theory.

But not unlike a pregnancy and birthing experience, scientific products of conception do not always result in something viable. Sometimes the idea is purposely aborted, or naturally miscarries, or simply dies much later in the scientific womb, resulting in a stillbirth. Occasionally a delivery becomes obstructed and requires a Caesarean section to be performed by a doctor. This occasionally happens in science. One person originates an idea, and someone else brings it to completion. Then, again, a real delivery may bring forth what appears to be a beautiful, healthy baby, only to discover later that a cardiac malformation will cut the life short unless corrected surgically. Sometimes this happens in science. What may appear at first to be a beautiful, new scientific idea will die unless major changes are made as, in real life a cardiac malformation is surgically corrected. In this way a scientific paradigm, or theory, is altered to fit the new data as more is learned about a given topic. However, some people, even scientists, become so enamored with their paradigms that they refuse to change or give them up just like some folks who still believe in a flat Earth. A preconceived idea must never be chosen over what is demonstrated to be real. To quote Niles Eldredge: "*Repeated failure to confirm predicted observations means we have to abandon an idea no matter how fondly we cherish it, or how earnestly we may wish to believe it is true.*"[1] Again, when a birthing experience produces a perfect baby, the birthing process is almost always painful. This is how it often is in science. Even when a new scientific idea finally becomes accepted in the scientific community, its initial delivery is often associated with much psychological pain and trauma borne by the originator. Occasionally, in life or science, twins, triplets, or quadruplets are delivered with what appears to be minimal effort or pain. Some scientific theories are even adopted for someone else to raise.

Through the course of history there have been many brilliant men who tried to explain natural phenomena. Unfortunately, at first they did not use testing methods, which would either prove or disprove their explanations of how something might look or work. As a result, many false explanations became impregnated in the minds of additional wise men and were handed down generation after generation with no one daring to question the truth of what they had been taught. This produced many false paradigms, some of which lasted for thousands of years. Webster's II College Dictionary defines a paradigm as "A set of assumptions, concepts, values, and practices that constitutes a way of viewing reality for the community that shares them, esp. in an intellectual discipline." A paradigm is similar to a scientific hypothesis or even a theory. Think of them as visualizing something before it's fully understood. The early Greeks proposed many mathematical and scientific paradigms, some of which have survived and some of which have been discarded.

Aristotle, (384 to 322 BCE) a Greek philosopher, taught that there were four Earthly elements: Earth, air, fire, and water. He believed that all celestial bodies were composed of a fifth element called *aither*. Aristotle considered *aither* to be a perfect substance. And because he believed that every heavenly body from the moon and outward away from Earth was composed of the perfect element *aither,* they therefore had to be perfect. He taught that they were perfectly round and traveled in perfect circles. In the arena of physics, Aristotle taught that heavier bodies would fall faster than lighter bodies as long as they had the same shape. About this time the dominant school of Greek mathematical astronomers taught that the Earth was stationary, located at the center of the universe, and that all heavenly bodies beyond the Earth were each attached to consecutively larger transparent, crystalline spheres that moved around the Earth, producing day and night. The moon was attached to the first crystalline sphere; the next contained the sun, followed by five consecutive spheres containing the five planets known to them. Altogether, these teachings prevailed for about 2,000 years, until Copernicus, Kepler, Galileo, and Newton made their debut on the scientific scene.

These paradigms were further bolstered by Claudius Ptolemy (150 CE). His mathematical calculations seemed to confirm the ancient Greek teachings.[2] Ptolemy's mathematical and astronomical writings, thirteen volumes in all, were preserved by the Arabs and became known as the Almagest, meaning "the greatest." In one volume, Ptolemy said that the Earth was stationary and the center of the universe (geocentrism). Like Aristotle, he thought that the moon, sun, and planets moved around the centrally placed Earth along with the stars. He believed the stars to be points of light attached to a concave dome. Ptolemy noted that the various planets moved at different speeds and sometimes seemed to stop and move backward against the backdrop of the distant stars. Ptolemy worked out an elaborate number of epicycles and equents, to mathematically predict where the planets would be at a given time. His paradigm lasted more than 1,200 years.[3]

Aristotle's teachings reached their acme about 1,500 years after his death when Thomas Aquinas (1225-1274) introduced them again into Western thought in 1266 in his Summa Theologica. He was so successful in this reintroduction of Aristotle and Ptolemy that their paradigms of "how the heavens go" and other concepts dominated Western teaching for about three centuries. In the minds of so-called educated elite and those in authority, this notion controlled their thinking so much that any alternate approach to this cosmology or other natural phenomena was considered unacceptable. It even could carry the penalty of death. This set the stage for the development of a deep antagonism between those with dogmatic paradigms and the newly emerging scientific community. [4]

Notwithstanding, Aristotle's cosmological conception was a stillbirth from its inception, but even though dead, was kept alive in the minds of very bright men for centuries. Ptolemy's math seemed somewhat resuscitative, causing Aristotle's paradigms to survive even longer, but to no avail. The truth about the "baby's" death had to wait for more than a millennium until Copernicus, Kepler, Galileo, and others performed an academic autopsy, which showed the causes of its demise.

Nicholaus Copernicus (1473-1543) was born and raised in Poland but as a young man went to Italy where he studied canon law and medicine. While a student at the University of Bologna, he studied astronomy as a sideline. His interest in it was stimulated while living in the home of a mathematics professor, Domenico Maria de Novara. The more Copernicus learned, the more he began to think that Aristotle and Ptolemy were wrong about the Earth being at the center of the universe. He began to think of the sun as the center (heliocentrism) and that the Earth circled around the sun. He was reluctant to tell many people about his beliefs because he might be arrested by the authorities. Eventually, however, when he was much older and living hundreds of miles away from Italy, he wrote a book titled *Revolutionibus Orbium Coelestium* (on the Revolutions of the Celestial Spheres), explaining his ideas. His book was published in Nuremberg, Germany, just before he died in 1543. In fact, it is believed that a copy of his newly published book was handed to him on his death bed only hours before he died. The birthing of the heliocentric paradigm took much of his lifetime. [5]

Johannes Kepler (1571-1630) was a German mathematician who for a time taught mathematics at University of Graz in Austria until he was driven out by Archduke Ferdinand, over a disagreement on a special matter. He fled with his wife and children back to Germany with two wagons of household goods. Later, he became the assistant to Tycho Brahe, a Danish astronomer, who used instruments other than telescopes to plot the courses of planets across the sky. All of Tycho Brahe's meticulous records, collected over many years of observations, fell into Kepler's hands when Tycho died about a year after the two began working together. From this data, Kepler was able to plot the path of the planet Mars in the sky. To his surprise, he found that Mars traveled in an elliptical path around the sun, with the sun at one focus of the ellipse. This surprised Kepler because Aristotle and Ptolemy had emphasized that the heavenly bodies traveled in perfect circles. His passion for finding the answers to these questions is demonstrated by the fact that it took him almost five years to complete the calculations on

Mars. This was because he did all of his calculations by hand using ink and quill. He had to repeat his calculations several times to insure he had made no mistakes. He may well have worked far into the night solving these problems by candlelight. Kepler discovered three laws of planetary motion. First, every planet follows an elliptical path around the sun. (An ellipse is like a circle with two centers, the sum of both radii at any given point on the ellipse remains constant). Second, as a planet goes around the sun, its speed varies so that a line from the sun to the planet sweeps over equal areas during equal times. Third, the time that it takes a planet to go around the sun once, when squared, is proportional to the cube of the mean distance from the sun. These laws are "Kepler's Three Laws of Planetary Motion." [6] All three of these scientific triplets were born viable, but only through many years of effort.

Just think, if Archduke Ferdinand had not forced Kepler out of Austria, he may never have found employment with Tycho Brahe. When Tycho died, most likely all of his valuable data would probably have died with him. We would never have heard of Kepler or his three laws of planetary motion. It was the third law that later became so critical for the delivery of Newton's universal law of gravitation.

Galileo Galilee was born in Italy February 15, 1564, the year of Shakespeare's birth, and in the same year that Michelangelo died. He lived until 1642, the year of Newton's birth.[7] Galileo, at age 17, entered the University of Pisa to study medicine. However, Galileo soon grew tired of studying Aristotle and Galen, enjoying the study of mathematics and physics instead. Much to the dismay of his professors, he soon began to attack the views of Aristotle on these subjects. At age 25, his reputation as a mathematician landed him a three-year appointment as professor of mathematics at the University of Pisa. He took advantage of this situation to study accelerated motion for the next three years. His studies led him to a clear understanding of acceleration and inertia. His contribution to physics at this time in his life was in the field of mechanics. His contract at Pisa was not extended at the end of the three years, undoubtedly because he antagonized his tradition-bound associates. It was during this time that he reportedly performed his famous experiment where he dropped two different sized weights from the leaning bell tower. Both weights hit the ground at the same time, which contradicted Aristotle's teaching that heavier weights fall faster than lighter weights. Whether it was the results of this experiment or other things that Galileo said or wrote, the status quo at Pisa was disturbed and he was forced to leave. However, shortly following the end of his contract at Pisa, he secured the appointment of professor of mathematics at the University of Padua where he continued his scientific investigations.

Nevertheless, it is Galileo's contribution to astronomy for which he is best remembered rather than his contributions to falling objects. Galileo was the first to use a telescope to study the heavens. He was a believer in the Copernican theory, and he made three telescopic discoveries, the last of which helped to confirm the heliocentric ideas of Copernicus and to also negate the geocentric ideas and other teachings of Aristotle and Ptolemy. Galileo's observations included three findings. First was the visualization of the mountains and valleys on the moon. Aristotle had taught that the moon was perfectly round and smooth. Second were the four moons "circling" Jupiter. Aristotle taught that all heavenly bodies circled the Earth. The third observation was that Venus passed through phases similar to the phases of Earth's moon. This could not happen if Venus circled the Earth as Aristotle and Ptolemy had taught. Poor Aristotle and Ptolemy, you would expect their paradigms to be in trouble by this time, but it was Galileo who was in trouble instead. It was the educated elite of Galileo's day who were determined to destroy not only these scientific triplets but their father as well. Considered one of the first of the modern physicists, Galileo confirmed his teaching with either experiments or observations, rather than depending upon what some ancient wise men had taught. Because he was so vocal about his findings, he incurred the ire of the authorities who thought of Aristotle's teachings as etched in stone. He was tried and, as a result, his last few years were spent in house arrest. [8] It was fortunate for him that he was not burned at the stake.

Isaac Newton was born prematurely on the morning of December 25, 1642. His father had died about three months before Isaac was born but Mrs. Newton married again soon after Newton's birth, leaving him to be raised by his elderly grandmother in a rural farmhouse. Newton was a sickly child, and some thought he would never reach manhood. Isolated from other children, he learned to play by himself, even after starting school. In grammar school, Isaac did not study very hard until he got into a fight one day with a fellow student whose grades were better than his. Isaac not only won the fight, but he also decided to defeat his opponent scholastically as well. This stimulus soon placed him first in his class. At home, he began to neglect farm chores for reading and building mechanical contraptions like windmills and water clocks. In fact, he even made kites in which he placed lanterns for flying at night. He also could draw very well, and he decorated his room with some of his own artistry. When he was about fifteen years old, and his mother again was living with him on the farm, she tried hard to make a farmer out of her studious son. But Newton showed no interest in farming, and she sent him back to school to prepare for entrance into London's Trinity College at Cambridge University, where he enrolled as a student in 1661.

Scholastically, he did not distinguish himself until he became a student of Professor Isaac Barrows, a mathematician who filled the Lucasian Chair of Mathematics at Cambridge in 1663. From then on, it seemed that Newton had found his niche as his extraordinary mathematical talent not only became evident, but also became his driving force. He received his Bachelor of Arts degree in 1665. However, the bubonic plague outbreak in London forced closure of Cambridge University for two years. This caused Newton to return to the Woolsthorpe farm. Now in his early twenties, Newton made some of his greatest discoveries during this time of "enforced idleness." He completed his theory of colored light, discovered the binomial theorem of algebra, and invented differential and integral calculus, all within a few months' time. However, because he did not publish his methods of calculus until much later, he was drawn into a controversy between himself and the German mathematician Gottfried Leibnitz who had discovered similar methods about the same time. Aside from the disagreements between himself and Leibnitz, this set of scientific and mathematical triplets seemed to be brought into the world without much pain or suffering. Also, during this two-year period, it is said that Newton conceived the idea of gravitation supposedly from seeing an apple fall from one of the trees on the farm. He completed the law of gravity much later after developing his Three Laws of Motion (another set of scientific triplets). The first law of motion states that an object will remain at rest unless moved by a force. The second law of motion states that an object's acceleration is equal to the net force on the object divided by its mass. From this is derived the famous equation force equals mass times acceleration or F=ma. The last or third one says, for every action there is an equal and opposite reaction. Important as these three laws of motion are, he is probably best remembered for his law of gravitation, which mathematically describes the attraction between two objects with a mass such as the earth or moon. Obviously, there was much more to his discovery of the universal law of gravitation than simply seeing an apple fall from the tree to the ground. It was from Kepler's Third Law of Planetary Motion that he made the deduction of the now well-known Inverse Square Attraction Law. It says that the force of attraction between any two bodies of matter is inversely proportional to the square of the distance between them and directly proportional to the mass of each body.

After the bubonic plague epidemic (also known as the black plague) was over, Newton returned to Cambridge University in 1667 and received a Master's Degree in 1668. The following year, he succeeded his teacher, Dr. Barrows, by becoming the chairman of the Lucasian Professor of Mathematics.

Newton mistakenly concluded the universe had to be infinite in size and static because gravity attracts every body of matter in the entire universe to every other body of matter. His reasoning went like this: If the universe had an edge, then matter located at the edge would start "falling," or in other words, it would be attracted to other matter toward the center of the universe. Thus, the universe would collapse on itself. If, however, the universe were infinite in size, there would be no edge and there would be no center. Therefore, matter could not be attracted toward a nonexistent center. With this idea of an infinitely large universe, all matter would be static in its space except for frictionless movement of the planets around the sun and the moons around the planets. Infinite and static were Newton's conception of the universe.[9] However, it was doomed to the fate of a post-partum death.

Newton died in 1727, leaving a tremendous legacy of basic discoveries. It is by following Newton's three laws of motion and universal law of gravity that allows us in the twenty first century to place artificial satellites in orbit along with their astronautical passengers. Modern computers programmed with Newton's rules of fluctions, or calculus as we know it today, plotted the trajectory that took astronauts to the moon and back.

Immanuel Kant (1724-1804), a famous Prussian/German philosopher, agreed with Newton's idea of an infinitely large, static universe and developed a cosmology of infinite time as well. With this, he set the stage for scientists that followed until the beginning of the twentieth century. An infinitely large and infinitely old universe was Kant's paradigm that he handed to those who followed.[10]

SUMMARY

1. A paradigm, like a scientific hypothesis, is a way of mentally visualizing something. It may be based at first on minimal information and gradually changed as more information becomes available.
2. In scientific terms, a theory may be similar to a hypothesis or a paradigm. However, it is based on more information than a hypothesis, and is thought to be correct. It may need changing as more data are obtained.
3. Aristotle taught that Earth was the center of the universe (geocentric) and that all other heavenly bodies, including the moon, sun, planets, and the stars were perfectly round and smooth, and circled Earth attached to transparent spheres.
4. Ptolemy held similar geocentric beliefs about the construction of the universe as did Aristotle. Even though the geocentric paradigm was false (except for our moon), Ptolemy was able to describe the movements of the heavenly bodies mathematically with considerable accuracy.
5. Copernicus (1473-1543) was probably the first person to propose that the Earth goes around the sun (heliocentric).
6. Kepler (1571-1630) discovered the three laws of planetary motion: (1) All planets travel in elliptical orbits around the sun, (2) a line drawn on the plane from the sun to the planet sweeps over equal areas during equal time,. (3) when the time for one elliptical trip around the sun by a given planet is squared, it will be proportional to the cube of the mean distance from the sun.
7. Galileo (1564-1642) discovered the mountains and valleys on the moon, four of Jupiter's moons, and the phases of Venus. All of these discoveries disproved paradigms of Aristotle. The last one helped to confirm the heliocentric paradigm. He also studied the acceleration of falling bodies.
8. Isaac Newton (1642-1727) described the reasons for colored light, discovered the binomial theorem of algebra, and invented calculus. He also discovered the three laws of motion and described the law of gravitation. Newton mistakenly concluded that the universe was infinite in size and static. According to him, it had no edge or center.
9. Immanuel Kant (1724-1804) believed that the universe was infinitely old.

Chapter 2

Darwin

The theory that Charles Robert Darwin gave to science (according to some authorities) has had more effect on Western society, as a whole, than any other scientific theory, including Einstein's Theory of Relativity.

Darwin was born February 12, 1809, the same day as Abraham Lincoln. He was the son of a wealthy physician who had made most of his fortune in real estate and other investments.[1] Unfortunately for Charles, his mother died when he was only eight and a half years of age.[2] A few months after his mother's death, his father sent him to a boarding school approximately a mile from his home. This proximity allowed Charles the freedom of frequent visits to his home after classes, although he spent nights in the dormitory at the school. A Unitarian minister named Samuel Butler, known as a strict disciplinarian, ran the Shrewsbury School that young Charles attended. Basically, it was a prep school for universities that emphasized Greek and Latin classics. Charles showed no interest whatsoever in this line of teaching.[3] However, about ten years of age, he began to show considerable interest in science. He started collecting insects and tried to categorize them into their various species. When he was about thirteen, he and his eighteen-year-old brother, Erasmus, or Eras as the family called him, pooled their funds and built a small chemistry lab in their backyard where they performed various experiments.[4]

Soon after, the two brothers were separated when Eras went to study medicine at Cambridge University near London. Meanwhile, Charles disdainfully anticipated several more years at Shrewsbury Boarding School.[5] He had at least two consolations, however. First, he had the run of the chemistry lab,[6] and second, he had a new-found preoccupation with hunting, especially birds.[7] In 1825, when Charles was 16, his father saw that his second son did not conform to the teaching methods of Shrewsbury and removed him from the school. Meanwhile, Eras had finished the first three years of medical school at Cambridge and needed to take his external hospital study year, which worked out to be at Edinburgh University.[8] With this in mind, their father, Dr. Darwin, decided to send 16-year-old Charles to Edinburgh to start medical school studies where he could room with Erasmus.[9] Charles took the usual courses for a first-year medical student, which included lectures in chemistry by Dr. Hope. Eras also took the same chemistry course.[10] The two brothers later began to skip classes, preferring to spend much time reading books from the library.[11, 12, 13] In addition, Charles frequently took refuge from his studies by walks into the country or along the coast looking for insects to add to his collection. He even found time to take a private class in taxidermy.[14] Somehow, both brothers managed to scrape by in school. In April of 1826, when Eras had completed his medical course, he left Edinburgh, leaving Charles alone at the university.[15] During his first year at medical school, Charles had decided that medicine was not for him when he witnessed two surgeries. One was performed on a child, and both were without anesthesia since none was available at the time. The sight was so grotesque to him that he had to leave.[16]

That summer, back home at the Mount (as his home in Shrewsbury was called), Charles managed to avoid telling his father of his dislike for medical school. There were long absences from home by paying frequent visits to friends. He hiked to Wales, took trips to Wedgwood's house, and spent many days hunting in earnest.[17] During that summer, according to a record found later, Charles shot a total of 177 hares, pheasants, and partridges.[18] It seems paradoxical that the medical student forced to leave the surgical theater because of the repugnance of human suffering, could delight in witnessing the animal suffering, and be its primary cause to boot.

When he returned to Edinburgh in the fall of 1826 to begin his second year of medical school, it was with no small amount of misgivings. His older brother was not there, which forced him out from under his shadow, causing him to seek new friends. Besides his usual studies, he took a class in natural history from Professor Robert Jameson. The syllabus of this class was to include such topics as zoology, botany, paleontology, geology, and mineralogy.[19] He also joined the Planian Society, which had a great impact on him. Most of the society was composed of undergraduate students. They met for reading scientific papers followed by discussions in an underground room at the university.[20] One of the members who had a profound influence on Charles, was Robert Grant. Though not a professor, he had graduated as a doctor of medicine from Edinburgh in 1814. It was Grant that got Darwin generally interested in marine biology. He also introduced Charles to evolutionary ideas by expounding the teachings of Lamarck. Later, Charles also studied the writings of Doctor Eras Darwin I, his grandfather, who had expressed similar ideas in his book called *Zoonomia.* [21, 22]

By the end of Charles's second year in medical school, he had decided that a medical career was not for him. He told his father he would quit.

His exasperated father, in a fit of anger, told Charles, "You care for nothing but shooting, dogs, and rat catching, and you will be a disgrace to yourself and all of your family."

If his son would not become a doctor in the footsteps of his father and grandfather, what could he do? He could not become a lawyer because of his ineptitude for Latin and Greek. A career in the armed forces or government did not appear promising. The Anglican Church was all that remained. His father sent Charles off to Cambridge. He arrived in January 1828, just before his nineteenth birthday, to prepare himself for the study of the ministry. Again, because he had forgotten practically all of his Latin and Greek, his father hired a private tutor to cram young Charles on these subjects before going to Cambridge.[23] Because of his late arrival in the middle of the academic year, Charles took up temporary quarters on Sidney Street. The following October he was able to secure quarters on the south side of the college. Many years before, William Paley occupied these quarters and in the court below, the poet John Milton had walked and talked.[24]

Despite this inspiring historic atmosphere, Darwin spent most of his time in anything but academic work. He put a great deal of effort into catching insects, especially beetles, which he continued to identify and label.[25] He even hired people to help him in this venture. Sometimes they tagged along with him, carrying various paraphernalia while he made the catches. Sometimes he sent them out on their own to catch insects for him.[26] During this time, one of his rich student friends anonymously presented him with a microscope. Many years later, Charles found out who his patron was. It turned out to be John Herbert.

Then Darwin wrote to Herbert, his friend and benefactor, referring to the gift: "I can hardly call to mind any event in my life that surprised and gratified me more."[27]

At this time, to illustrate just how serious Darwin was in his collection of insects, specifically beetles, he told a story in his autobiography. "One day, on tearing off some old bark, I saw two rare beetles and seized one in each hand. Then I saw a third, a new kind, that I could not bear to lose, so I popped the one I held in my right hand into my mouth. Alas, it injected some intensely acrid fluid, which burnt my tongue, and I was forced to spit the beetle out. That beetle was lost as well as the third one."[28]

When he was not searching for beetles, he would, likely as not, be in the woods shooting birds. He passionately pursued this sport in summer and autumn when birds were plentiful. In the off-season, Darwin spent considerable time studying books on the sport of hunting.[29]

During his years at Cambridge, Darwin became quite attached to his Uncle Josiah Wedgwood. Uncle Jos, as he was called, liked to hunt as much as Darwin. He owned a large estate, which made an ideal setting to carry out hunting escapades. Uncle Jos had four sons, none of whom took to this particular sport. Darwin was drawn to this uncle on his mother's side, and vice versa. During the summers between sessions at Cambridge, invitations to Darwin to join his uncle during the partridge or pheasant season were not uncommon.[30]

Frequently, a gentleman named William Owen invited Darwin to his estate. Not only did he like to hunt, but he was also the father of two beautiful and eligible daughters. Fanny, the prettier of the two, caught Darwin's fancy for a time, and romance seemed to flourish when her father invited Darwin to his estate to hunt. This romance ended when Darwin left to set sail on the *Beagle*, the ship that was to circumnavigate the globe.[31]

We must understand that college tutors did most of the teaching at Cambridge University. They helped students prepare for examinations. Attending the lectures of professors was not a requirement. In fact, students avoided them when possible. It was only with the help of a dedicated tutor, who encouraged Darwin to cram for examinations that he got through. To everyone's surprise, including Darwin's, he graduated tenth in a class of one hundred and seventy-eight.[32]

Though progressing toward a Bachelor of Arts degree, which was supposedly to prepare him for the ministry in the Anglican Church, Darwin spent much of his scholastic time on anything but pursuing that goal. As already noted, hunting, entomology, and socializing with friends seemed to occupy most of his time. It almost seemed unnatural for one professor, John Stevens Henslow, to draw Darwin's attention. Later, Darwin credited him with influencing his career more than any other. Professor Henslow was different in two ways from other professors. First, he was young (only twenty-six years old) when appointed a professor of mineralogy. Shortly afterward, he became a professor of botany. Second, he took an active part in teaching. His lectures were very popular. Professor Henslow was thirteen years older than Darwin, but the two seemed destined to become the best of friends. He supplemented his lectures with practical instruction, not only in recognizing different plants, but in their dissection as well. In addition, there were excursions into the field, sometimes on foot or stagecoach or even a river barge. However, it was the Friday evening intellectual soirées that Henslow held at his home, where students and teacher became acquainted in a less formal atmosphere. There were usually no more than ten or fifteen students, and occasionally a famous guest would be present. Originally, Darwin's invitation to these Friday night gatherings came from his cousin, William Fox. At first, Darwin did not attend these meetings regularly. It was in his last two terms at Cambridge that Henslow and Darwin's friendship seemed to mature, with Henslow being a mentor for his young pupil.[33]

Another professor at Cambridge that influenced Darwin was Professor Adam Sedgwick, a professor of geology. Toward the end of his last year at the university, Darwin dreamed of an excursion to the Canary Islands with friends and Henslow. Although this excursion never happened, Henslow told Darwin that he could not hope to trek over the volcanoes of the Canaries without a basic knowledge of geology. It was Henslow who initiated Darwin's studying this subject. Later, he arranged for Darwin to accompany Professor Adam Sedgwick on a geological field excursion that he was planning for the summer vacation of 1831. So that summer, after graduation, Darwin spent weeks in the field studying practical geology under the private tutelage of Sedgwick.[34]

Darwin arrived at home August 29, to find a very surprising letter waiting for him. The letter from Professor Henslow changed Darwin's life forever. It was an invitation to be a naturalist on board the ship *Beagle*, which was to chart the waters on the southern tip of South America and then circumnavigate the globe. Although they applied the term naturalist to his title, Darwin's main function was to be a companion to Captain FitzRoy, who was in charge of the mission. Except for the unpaid position of naturalist companion, the British Navy bore the cost of the mission. The passage fee alone was £500.[35]

It was Darwin's intention, now that he had a BA degree, to return to Cambridge that fall to continue his studies for the clergy. The invitation elated him, but when he revealed the contents of the letter to his father, Darwin was met with a wall of negatives. Without his father's psychological and financial backing, he faced losing this opportunity. As a consequence, the next morning (August 30) Darwin glumly wrote to Henslow, declining the offer. He already had planned to go partridge hunting with Uncle Jos on the first day of September. After writing the letter, he left for his uncle's estate, but not before his father had some second thoughts.

Before leaving, his father said that if Charles could find any sane person with sound judgment who thought the trip would be a good idea, Dr. Darwin not only would give his consent but also finance the journey. Darwin knew his father was speaking of Uncle Jos. To his great surprise and relief, the whole Wedgwood family, including Uncle Jos, thought the proposed trip was a wonderful opportunity. The next day, Darwin and his uncle wrote letters to Charles's father giving reasons Charles should go on the trip. These were sent September 1, after which the two men went hunting. However, neither man had their heart in it. Later, about 10 o'clock that morning, Uncle Jos accompanied his nephew back to the Darwin home to meet Dr. Darwin face to face. It was a needless journey as the letters had already done their job. Dr. Darwin not only would give his blessing, but also finance the trip. Charles was to go. So Charles had to compose another letter to Henslow, recanting the first, stating that he would be able to go after all.[36]

That fall, instead of returning to Cambridge to further his studies in the ministry, Darwin spent his time preparing for the trip. Because he was to be a companion to Captain FitzRoy, it was of necessity that he meet with him before leaving. The two seemed to get along quite well at their first meeting. Darwin was then twenty-two, and Captain FitzRoy was twenty-six. [37, 38]

The next few weeks were a flurry of preparations for the proposed two-year trip. These included the purchase of at least one large rifle, a telescope, other scientific measuring devices, writing materials,[39] and the stowing away of books. Not the least of these books was Charles Lyell's great work, *The Principles of Geology*, the first volume published in 1830. The *Beagle* finally set sail on December 27, 1831,[40] after being blown back by contrary winds on two previous attempts to leave England. The second volume of Lyell's work was published after the *Beagle* left England; someone forwarded it to Darwin while still on the voyage.[41] Unknown to the captain and his crew on the *Beagle*, they would not return until October 2, 1836, almost five years later. During the trip, Darwin made meticulous notes, collected specimens of rocks and fossils, and hunted various birds and other creatures that he took aboard the *Beagle* as specimens to be studied later.[42] His previous training in taxidermy proved to be a great asset. [43] This was especially true on both the Eastern and Western coasts of South America,[44] where he spent considerable time on land. However, the apex of his scientific observations occurred when the *Beagle* visited the Galapagos Islands.[45] Here was pristine flora and fauna, the likes of which he had never seen before.

The Galapagos Archipelago is situated about six hundred miles west of Ecuador. When Darwin visited these islands, he was struck by the apparent differences between the various land birds that he saw there, compared with similar ones he had seen on the continent of South America, 600 miles to the east. Various finches living there especially attracted his attention. They bear the name Darwin finches to this day. He reasoned that the Galapagos Islands were of volcanic origin; that is, they had raised up from the floor of the ocean as the lava broke through from deep in the Earth. That meant that these islands were younger than the South America continent. Volcanic soil began to form many years after the lava had cooled and started to break down. Later, seeds carried there by water, wind, or seabirds, began to vegetate the islands. Later, insects colonized the plants. He speculated that later yet, a few land birds settled there after having been blown off course by a storm. Darwin's finches, he postulated, were among these early settlers. As their descendants adapted to their new surroundings, they began to develop various characteristics that differed from their progenitors on the mainland. These changes were found especially in the shape of their beaks. Darwin noted that some of the finches developed thick, heavy bills for cracking and eating seeds, and some developed thinner bills to dig insects underneath bark. It was easy for him to collect animal specimens because the animals had no fear of man and would not run as he approached. Before the *Beagle* left these islands, heading west, Darwin stowed all of his specimens to be studied later.[46]

The *Beagle* continued from the Galapagos to visit the South Sea Islands, including Tahiti, New Zealand, Tasmania, and the Australian continent. Continuing west from Australia, they visited Keeling Islands, the Mauritius, Cape of Good Hope, Ascension Island, and then back to South America.[47] They then traveled north, arriving back in England at Falmouth on October 2, 1836. Darwin was so eager to get home that he left ship the next morning. Driving his horses as hard as he could, he arrived home late on October 4. Everyone was asleep. He quietly let himself in and went to his room. His family knew nothing of his arrival until he casually walked in to breakfast the next morning.[48]

After Darwin returned to England, he studied the various specimens he had collected. As he went over all of his meticulously written notes, he began to think that all of the finches on the Galapagos Islands had descended from the original few birds that had arrived there by accident. From this, he saw descent with modification, which he later called variation under nature. These changes seemed to have improved the ability of the finches to survive on those volcanic islands. But one question remained: what was the driving force that made these changes take place?[49] You will learn in subsequent chapters about natural selection. Darwin believed that natural selection was that driving force which some folks refer to as survival of the fittest.

Since this chapter is simply a short biography of Darwin's life, no summary is provided.

CHAPTER 3

Biology 101 DNA, Life's Pattern

What is life? This may seem like a trivial question, but actually scientists have not been able to agree on a definition of life that they all accept. Some may say, "When I see it, I will know it," but this doesn't really produce a definition. Yet the most casual observer will almost always be able to tell when something is alive or dead. Gerald Joyce, a biologist at Scripps Research Institute, came up with a so-called "working definition" for life. Joyce said, "Life is a self-sustained chemical system capable of undergoing Darwinian evolution[1]." Everyone knows that all living things extract energy from their food and use it to power life's machinery. This process allows every living organism to flourish and reproduce. From these obvious characteristics of life, one would think that a definition acceptable to all scientists would be forthcoming, but one would be wrong.

Darwin noted that in the same species, some variations occurred in almost all life forms between each succeeding generation. He concluded that if any of these slight variations proved beneficial, its owner would be able to compete better than other biota for food and space and leave more descendents like itself. He thought, if this process happened repeatedly generation after generation without significant limitations, these beneficial traits would gradually accumulate over very long periods of time, thereby producing new species. On the other hand, deleterious traits would have the opposite effect, making extinction a possibility for any living form possessing them[2]. Fossils are proof that extinction has occurred in the past, which gives some credence to Darwin's proposal.

Looking for scientific evidence placing boundaries or limitations on evolutionary theory will be the core subject of this book. Many scientists believe this effort to be counterproductive and contrary to understanding biology itself. However, before evolutionary theory, along with all its implications, can be properly understood, at least a brief foray into basic biology must be made. To add a little zest to this study, it will be punctuated here and there with a few stories about the people who made some of the most momentous biological discoveries. One of these was the structure of the DNA molecule that stores the patterns of life. So here we go. Hang on for dear life, whatever life means, 'cause truth always bears inspection!

James Watson received his Ph.D. in biochemistry from the University of Indiana in 1950. After graduation, when only 22, he began doing post-doctorate research at the Cavendish Laboratory at the University of Cambridge, in England. He teamed up with a British biophysicist, Francis Crick, who had obtained a B.S. degree in physics from University College, London. Crick was about 12 years older than Watson. For nearly three years, the two worked to determine the chemical structure of the molecule that stores genetic information. This molecule turned out to be DNA, the abbreviation for deoxyribonucleic acid. DNA transmits its genetic information from one generation to the next in all living organisms. Crick and Watson's success was reached in April, 1953, the same year that Crick received his Ph.D. from research on the structure of the hemoglobin molecule. However, their work had been coaxed along with the help of another British biophysicist, Maurice Wilkins, who used X-ray diffraction techniques for this research. In 1962, about nine years after completing their research, Wilkins, Watson, and Crick shared the Nobel Prize in Physiology or Medicine for discovering the structure of the DNA molecule.

This seemed well and good at the time, but it is now known that a woman involved in the research never received any credit for this seminal accomplishment. Her name was Rosalind Elsie Franklin. Educated at Cambridge, Rosalind received a Ph.D. degree in physical chemistry in 1945, at the age of 25. Then starting in 1947, she spent three years in France learning how to photograph molecules using X-ray diffraction techniques. Returning to England in 1951, she began the ground breaking research to find the structure of DNA. Her X-ray diffraction photographs of the DNA molecule showed it to be a double helix. DNA is a mega-molecule that if visible, would look like a very long zipper, the sides of which could be seen as being composed of sugar and phosphorous, while the teeth of the zipper were made of four different kinds of chemical bases united in the center. Twist the double stranded "zipper" shaped molecule round and round on itself like a peppermint and a better representation is obtained. It was Franklin who identified the location of the phosphate group and sugar forming the sides or strands of the DNA molecule.

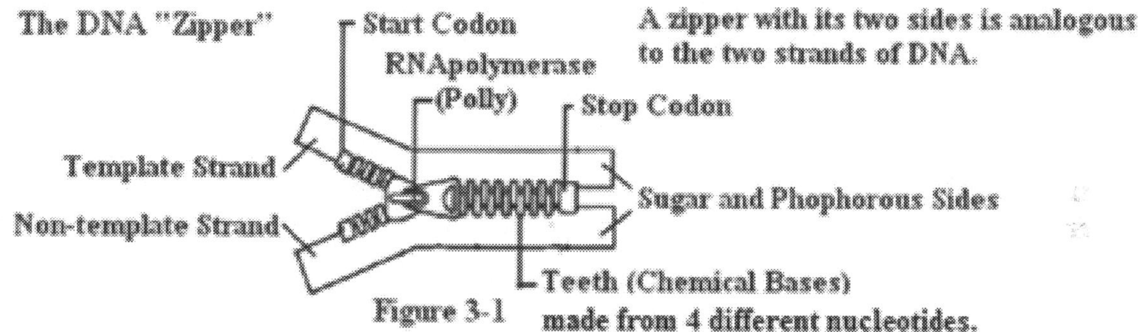

Figure 3-1

A colleague at Kings College, where Franklin was doing her research, was Maurice Wilkins. They did not get along well. Without Franklin's permission, Wilkins shared some of her work with Crick and Watson at Cambridge. When *Nature,* one of the world's most prestigious scientific journals, published Crick and Watson's article on the structure of DNA in the April 25, 1953 edition, Franklin apparently did not know what Wilkins had done with her work, even though she and Wilkins had a companion article in the same issue of *Nature.* Several years later at the age of 38, about three years before the Nobel Prize was awarded to the other three, Rosalind Franklin died of cancer in 1958. Even though Crick, Watson, Wilkins, and Franklin were the first to describe what DNA looked like,[3] the genetic code that it contained took about another decade for scientists to decipher.

Because DNA is the only place where the changes required by evolution can take place[4], it is simpler for the purpose of this book to divide all biota in two groups, based on the DNA of each. The two groups are **prokaryotes** and **eukaryotes[5].** The differences between them are defined by how their DNA is stored and whether or not it is divided into sections. The **prokaryotes** are all **single-celled** bacteria. Their DNA is double stranded like the two sides of a zipper, but also forms a circle with no beginning or end. It lays "naked," so to speak, in the cellular liquid known as cytoplasm. All **eukaryotes** store their DNA in a nucleus in multiple linear paired sections, each with a beginning and an end. Some are single-celled, such as yeast or malaria, but most are multi-celled like moss, magnolias, monkeys, or men[6].

As noted above, every tooth in each side of the double stranded DNA "zipper" can be considered analogous to a chemical base or a nucleotide. There are four kinds. Their respective names are **adenine, thymine, guanine,** and **cytosine**[7]. No need to learn these strange names, because each will be abbreviated by the capitalized first letter of its name. Adenine and thymine will be designated by the upper case letters (A) and (T). The same applies to guanine and cytosine, being represented by the upper case letters of (G) and (C). When (A) is located on one strand, it will always be connected to (T) on the adjacent strand forming a pair. (G) and (C) will also form another pair, always on adjacent strands. These two duets are called **base pairs.**

In your mind's eye, the base pairs can be visualized in DNA as (A)><(T) and (G)><(C). The zipper tooth of (A) fits (T)'s and (G)'s zipper tooth fits (C)'s. The DNA

DNA

Non-Template or Sense Strand →   ← Template or Anti-sense Strand

(A)><(T)
1. (T)><(A)
(G)><(C)
(G)><(C)
2. (T)><(A)
(T)><(A)
(G)><(C)
3. (C)><(G)
(C)><(G)
(A)><(T)
4. (C)><(G)
(C)><(G)
(A)><(T)
5. (G)><(C)
(C)><(G)
(C)><(G)
6. (A)><(T)
(A)><(T)
(C)><(G)
7. (A)><(T)
(T)><(A)
(T)><(A)
8. (G)><(C)
(A)><(T) Figure 3-2

pairs always appear together. It doesn't matter on which DNA strand (A) is located, as long as its mate, (T), is located across from it on the adjacent strand. The same principle applies to the (G) and (C). Each letter in each pair is said to be complimentary to the other. (A) is complimentary to (T), (G) is complimentary to (C), and vice versa. DNA is composed of millions of complimentary base pairs connected to each other in a long line. The genetic instructions in the DNA are "spelled out," coded by specific sequences in which the base pairs appear in that line.

Now move your mind over from the zipper analogy to actual DNA. On the opposite side of this page, look at the ultra-short column of 24 base pairs, which is not circular nor is its length millions of base pairs long. However, deciphering this short straight column of DNA will be the same as deciphering a very long circular one.

Notice that the letters forming each of the two strands of the DNA column are labeled as non-template and template strands and that the column itself has been divided into eight sections composed of three base pairs each. From top to bottom, the letters in the template strand on the right side of the DNA form different combinations of triplet sets. Using the four letters A, T, G, and C, there can only be 64 possible triplet sets.

Learning about RNA polymerase[8], comes next. It is a microscopic cellular machine, which we'll call, Polly, the stenographer for short. She transcribes the triplets in the DNA template strand into a long "ticker tape" single strand of messenger RNA or mRNA. It has a fifth base, uracil, designated by the uppercase letter (U). Each set of triplets in the mRNA forms a **codon**.

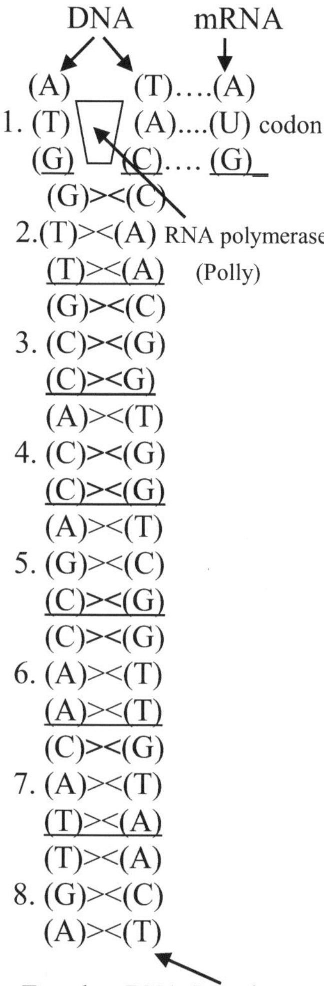

RNA polymerase (Polly) types a single strand of mRNA next to the template strand of DNA as she begins unzipping the two strands. Each base she types in the single strand of mRNA is complimentary to the corresp-onding base in the template strand of DNA. However, in the mRNA strand, (U) is considered complimentary to (A) in the DNA strand. No (T) is allowed in the mRNA single strand.

Figure 3-3

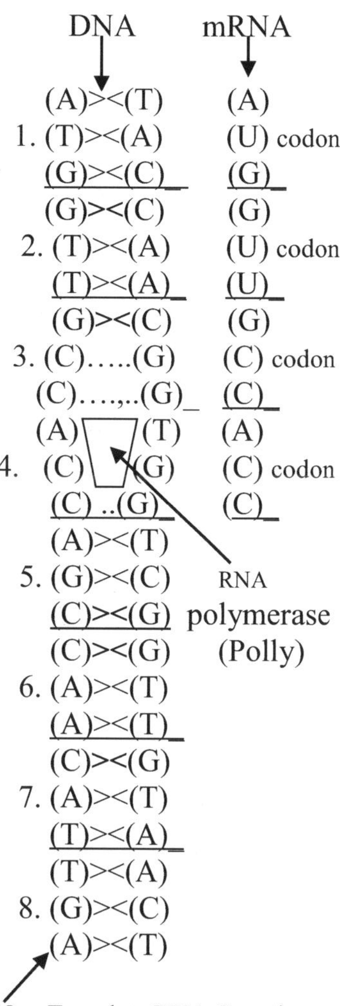

As Polly unzips the two DNA strands apart, she descends between them typing the mRNA from the template strand. The two strands of DNA reunite again behind her as she descends. Notice the different triplet permutations of (A), (U), (G), & (C) that she transcribes into the mRNA each time she moves past three base pairs in the DNA. These triplets are called codons, about which you will soon learn.

Figure 3-4

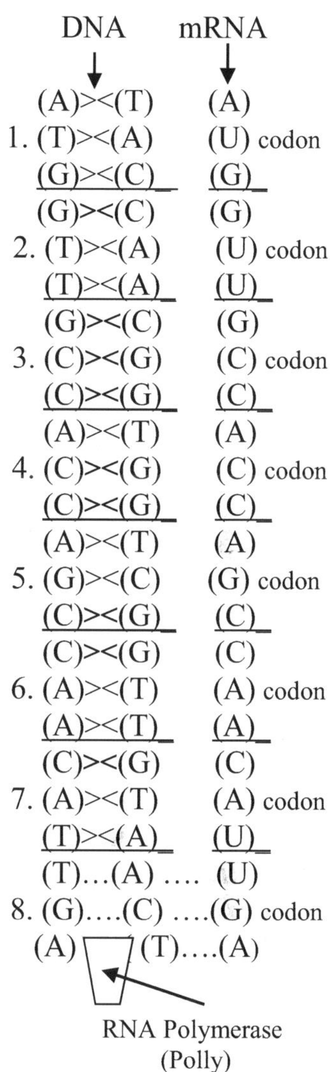

Polly has reached the end of the DNA column and is exiting. The process of making mRNA from the template strand of DNA is called transcription. The mRNA strand is sent to the ribosome[9] machine, we'll call "Mr. Ribo," who translates it into a protein specific to the cell where it resides. There are millions of different kinds of proteins.

Figure 3-5

The smallest unit of life is the **cell,** inside of which with rare exceptions, lives a bundle of DNA. This complete DNA bundle is called the genome of the cell and is analogous to an entire book of instructions for making the cell or needed repair parts. It is composed of **chromosomes**[10] that are divided into many **genes**[11]**.** Think of a chromosome as a chapter in the book and a gene as one of many paragraphs in a chapter. There can be thousands of gene paragraphs in one chromosome chapter and the number of chapters in a given genetic book can vary from as few as one on up to more than 100. The message in the gene is "spelled out" in triplet codons in mRNA.

There can be only 64 different combinations or permutations of three of the four letters A, T, G and U. These form triplet base pairs called codons[12]. Most genes contain the information for making one cellular part "spelled out" in different sequences of the 64 codons. Every gene begins with the codon AUC that's like a capital letter at the beginning of a written sentence. They all end with one of three stop codons UAA, UAG, or UGA that are similar to periods. In between the start codon and the one at the end are codons that contain the coded instructions for making the desired part. Every cell is made of many parts and most of these parts are proteins.

The building blocks of proteins are twenty different kinds of biological amino acids. Every kind of protein is made from a specific sequence combination of some or all of the twenty different kinds. Many amino acids attached to each other, end to end, in a specific sequence, form a long line like a freight train with 100 or more cars. Unlike a train, the line folds on itself making various three dimensional parts, each with a particular cellular function[13]. By themselves, amino acids are not dangerous or powerful like sulfuric acid. A specific sequence combination of amino acids used to make a protein found in heart muscle can be considered very friendly. However, a different combination forming snake venom is obviously very dangerous.

As noted on the previous page, after the single strand of mRNA has been typed up by Polly, the stenographer, it is then sent to another microscopic machine, the ribosome, that we dubbed "Mr. Ribo." Being bilingual, he is able to translate each successive mRNA codon into an amino acid, each one of which he attaches end to end making a long amino acid train called a protein.

When a repair part for the cell is needed, "Polly" goes to the gene located somewhere in the DNA that contains the desired pattern and types up an mRNA "tape," one base at a time. Each base in the mRNA is complimentary to the corresponding base in the template strand of the DNA. The "tape" is then sent to "Mr. Ribo," who reads it one triplet codon at a time, translating each into one amino acid, following the sequence of codons in the mRNA, which in turn were derived from the DNA. Information in a gene for making one part is first transcribed into mRNA, which is then translated into protein. With no thinking on our part, this process is occurring hundreds of thousands of times per second in each of the billions of cells in our bodies.

A couple of pages ahead, you will see a Genetic Decoder, which displays the 64 possible triplet permutations of the letters A, T, G, and C that compose the DNA and the 64 possible triplet permutations of the letters A, U, G, and C that form the 64 codons of mRNA. Each codon is a code for one of the 20 biological amino acids. On the next page, look at the three examples of codons to learn about reading the Genetic Decoder.

A Brief Description of How to Read the Genetic Decoder

The bases across from each other in each strand of DNA are complimentary to each other. The bases in the single strand of mRNA are also complimentary to each base in the template strand of DNA from which they were transcribed, except (A), in the template strand of DNA, is always transcribed as (U), in the mRNA.

DNA		mRNA	Amino Acid
1	**2**	**3**	**4**
(A) ---	**(T) ---**	**(A)**	
(T) ---	**(A) ---**	**(U)**	**Methionine**
(G) ---	**(C) ---**	**(G)**	
(G) ---	**(C) ---**	**(G)**	
(T) ---	**(A) ---**	**(U)**	**Valine**
(T) ---	**(A) ---**	**(U)**	
(G) ---	**(C) ---**	**(G)**	
(C) ---	**(G) ---**	**(C)**	**Alanine**
(C) ---	**(G) ---**	**(C)**	

**Column 1 represents the non-template strand or sense strand of DNA.
Column 2 represents the template strand or antisense strand of DNA.
Column 3 represents the single strand of mRNA transcribed from the DNA.
Each triplet of bases in column 3 represents one codon that specifies one specific amino acid in column 4.**

Table 3-1

Now peek at the Genetic Decoder on the next page. You can see that there are four large vertical columns labeled at the bottom with Roman numerals. Along the left hand side, numbers 1 to 16 are displayed, which together with the four vertical columns, show the 64 possible permutations of the bases composing the DNA and mRNA. At the top of each of the four vertical columns are numbers 1 and 2 that denote the two strands of DNA and 3 that denotes the single strand of mRNA. The numbers at the top of the decoder are the same as those noted above. If you look at the list of 64 codons, you will see that multiple codons can specify for the same amino acid, but no codon can specify for more than one.

Notice 8 rectangles in the decoder numbered 1 to 8, inside of which are the letters denoting the DNA and mRNA. To the right of each rectangle is displayed the amino acid, which is coded by the complimentary triplets inside. The 8 rectangles are numbered in the same sequence as the triplet codons forming the mRNA in the far right hand column, shown two pages back. Using the Genetic Decoder, translate the mRNA inside each rectangle into its amino acid and then into protein. See if you decode the same protein as shown on the page following the decoder.

The Genetic Decoder for DNA, mRNA, and Amino Acids

1 - Non-Templet Strand or Sense Strand
2 - Templet Strand or Antisense Strand
3 - Single Strand of mRNA

	DNA mRNA 1 2 3	AMINO ACID	DNA mRNA 1 2 3	AMINO ACID	DNA mRNA 1 2 3	AMINO ACID	DNA mRNA 1 2 3	AMINO ACID
1	T A U T A U T A U	Phenylalanine	T A U C G C T A U	Serine	T A U A T A T A U	Tyrosine	T A U G C G T A U	Cysteine
2	T A U T A U C G C	Phenylalanine	T A U C G C C G C	Serine	T A U A T A C G C	Tyrosine	T A U G C G C G C	Cysteine
3	T A U T A U A T A	Leucine	T A U C G C A T A	Serine	T A U A T A A T A	Stop	T A U G C G A T A	Stop 8
4	T A U T A U G C G	Leucine	T A U C G C G C G	Serine	T A U A T A G C G	Stop	T A U G C G G C G	Tryptophan
5	C G C T A U T A U	Leucine	C G C C G C T A U	Proline	C G C A T A T A U	Histidine 7	C G C G C G T A U	Arginine
6	C G C T A U C G C	Leucine	C G C C G C C G C	Proline	C G C A T A C G C	Histidine	C G C G C G C G C	Arginine
7	C G C T A U A T A	Leucine	C G C C G C A T A	Proline	C G C A T A A T A	Glutamine 6	C G C G C G A T A	Arginine
8	C G C T A U G C G	Leucine	C G C C G C G C G	Proline	C G C A T A G C G	Glutamine	C G C G C G G C G	Arginine
9	A T A T A U T A U	Isoleucine	A T A C G C T A U	Threonine	A T A A T A T A U	Asparagine	A T A G C G T A U	Serine
10	A T A T A U C G C	Isoleucine	A T A C G C C G C	Threonine 4	A T A A T A C G C	Asparagine	A T A G C G C G C	Serine 5
11	A T A T A U A T A	Isoleucine	A T A C G C A T A	Threonine	A T A A T A A T A	Lysine	A T A G C G A T A	Arginine
12	A T A T A U G C G	Methionine 1 Start	A T A C G C G C G	Threonine	A T A A T A G C G	Lysine	A T A G C G G C G	Arginine
13	G C G T A U T A U	Valine 2	G C G C G C T A U	Alanine	G C G A T A T A U	Aspartic Acid	G C G G C G T A U	Glycine
14	G C G T A U C G C	Valine	G C G C G C C G C	Alanine 3	G C G A T A C G C	Aspartic Acid	G C G G C G C G C	Glycine
15	G C G T A U A T A	Valine	G C G C G C A T A	Alanine	G C G A T A A T A	Glutamic Acid	G C G G C G A T A	Glycine
16	G C G T A U G C G	Valine	G C G C G C G C G	Alanine	G C G A T A G C G	Glutamic Acid	G C G G C G G C G	Glycine
	I		II		III		IV	

Upper case letters represent individual nucleotides

T = Thymine   A = Adenine   U = Uracil   C = Cytosine   G = Guanine

Table 3 - 2

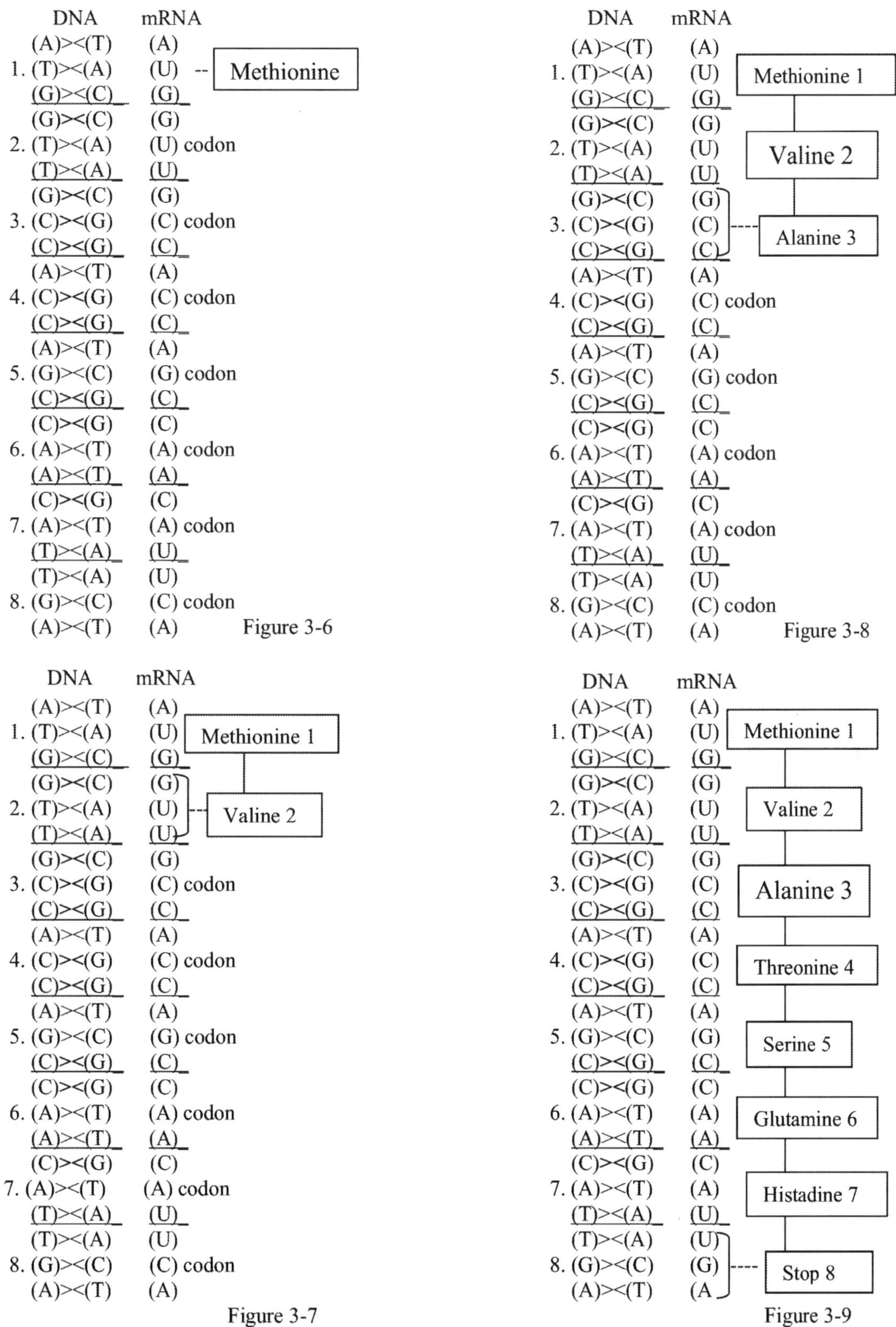

Figure 3-6

Figure 3-8

Figure 3-7

Figure 3-9

Summery

1. Collectively, Scientist have not been able to produce a definition of life that satisfies everyone.
2. Darwin noted slight changes between one generation and the next of the same species.
3. Darwin thought that if some of these changes were beneficial to its possessor that it would give that individual and increased chance to compete for food and space. If on the other hand the changes were deleterious extinction would ensue. (Although discovered many years after his time, these changes that Darwin noted, come from various combinations of DNA or mutations in the DNA).
4. Crick, Watson, Wilkins, and Russell elucidated the structure of DNA.
5. Since DNA is the molecule of inheritance, it is the only place where evolution can occur.
6. Prokaryotes (bacteria) store their DNA in one circular chromosome.
7. Eukaryotes (all other kinds of life than bacteria) store their DNA in multiple linear chromosomes in a nucleus.
8. DNA is made up millions of the two different kinds of base pairs stacked on top of each other, adenine and thymine: (A) and (T), and guanine and cytosine: (G) and (C).
9. mRNA is a single strand of bases (G), (C), (A), and (U). (U) in the mRNA is substituted for (T) in the DNA.
10. There are only 64 possible different triplet sets of the four bases (G), (C), (A), and (U). Each one of these different triplet sets of these four bases in the mRNA stands for a codon.
11. Each codon in the mRNA codes for only one of the 20 kinds of biological amino acids. However, more than one codon can code for the same amino acid.
12. Proteins are made up of different combinations of some or all of the 20 biological amino acids, which can vary from as few as 100 up to thousands.
13. Different proteins form different parts of all species of life.
14. The DNA in a given gene contains the information for making a specific protein or part of a protein that forms a specific part of a cell.
15. When a specific part is needed by a given cell for growth or repair, RNA polymerase goes to that gene and transcribes the single strand of mRNA from the message found in the DNA.
16. The mRNA is then taken to the ribosome that translates the message in the mRNA into a specific protein, which is the part needed by this cell.

CHAPTER 4

Biology 102 DNA Replication

The Basis of Life's Reproduction

Since all life, as we know it, dies at some time during its existence, it is imperative that all species of biota must reproduce themselves in order for life as a whole to be maintained and each individual species in particular to survive. Without considering the minimum number of individuals needed for any given species to maintain itself, this chapter will be confined basically to how the genomic pattern in the DNA in each species replicates; because without DNA replication no species could survive. The previous chapter was devoted to how parts needed for growth or repairs of an individual cell were made from instructions located in the DNA. This chapter will be an abbreviated attempt to explain how the entire DNA genome is replicated.

In order for each life form to reproduce itself, the pattern in the DNA that contains this information for making all the parts of an individual must be replicated and passed on to the offspring. As noted in the previous chapter, the genome containing the DNA of every species is made up of chromosomes, each of which in turn is composed of many genes. Except for genes that contain the pattern for making the various RNA's, almost 100% of them contain the information for building an individual part of a given organism. When all the parts are put together according to the pattern found in the entire genome, a new individual is produced very similar and in many cases exactly like its predecessors. It is very important to understand how the entire genomic pattern located in the DNA can replicate itself in order to preserve the species and project a copy of the genome in its entirety to a succeeding generation of biota.

Consequently, the instructions for making a new individual of any given species must be derived first from reproduction of its DNA. In order for this to occur, the double strands of the DNA in a replicating cell must first be split apart so that each of the two individual strands can act as a template on which to build two new complimentary strands of DNA, each exactly like its predecessor. When this process is finished, the cell temporarily contains two new genomes or two complete sets of DNA, each containing the pattern needed to construct two individuals of that spices. The cell can then divide in two by simple cell division with each half receiving a new double stranded set of DNA. This is what occurs in the reproduction of all cells. When completed, this process yields two new individual cells, each like the original, both containing a full complement of DNA. The diagrams that follow will demonstrate in multiple steps a simple cell division in progress.

Prokaryotes reproduce only by duplicating the DNA followed by simple cell division, where as eukaryotic DNA is first replicated by a process called mitosis and then the cell divides. This will be described and illustrated with diagrams in the next chapter. However, when prokaryotic replication is understood, it is relatively easy to transfer one's understanding over to the reproduction of eukaryotes. Therefore, we'll start with prokaryotes.

All prokaryotes have only one chromosome made up of a circular chain of DNA. If visible, it would look similar to a wadded up rubber band that has no beginning or end. When untangled, the chromosome like a rubber band, would assume an oval or circular shape, forming a continuous genomic ribbon. Many genes composed of base pairs attached to each other, make up the chromosome. As an example, the single chromosome of a bacterium E. coli is known to contain more than 4,500,000 base pairs and about 4400 genes in its genomic ribbon. However, to use less space inside a bacterium, the circle of genes forming the chromosome is all wadded up.

Learning how the replication of the genomic DNA of prokaryotes occurs, helps to understand how the pattern for each species is maintained and passed on to succeeding generations. The diagrams, which follow, will show what happens in only a short section of DNA reproduction. However, this will demonstrate what occurs during DNA replication of the entire genome.

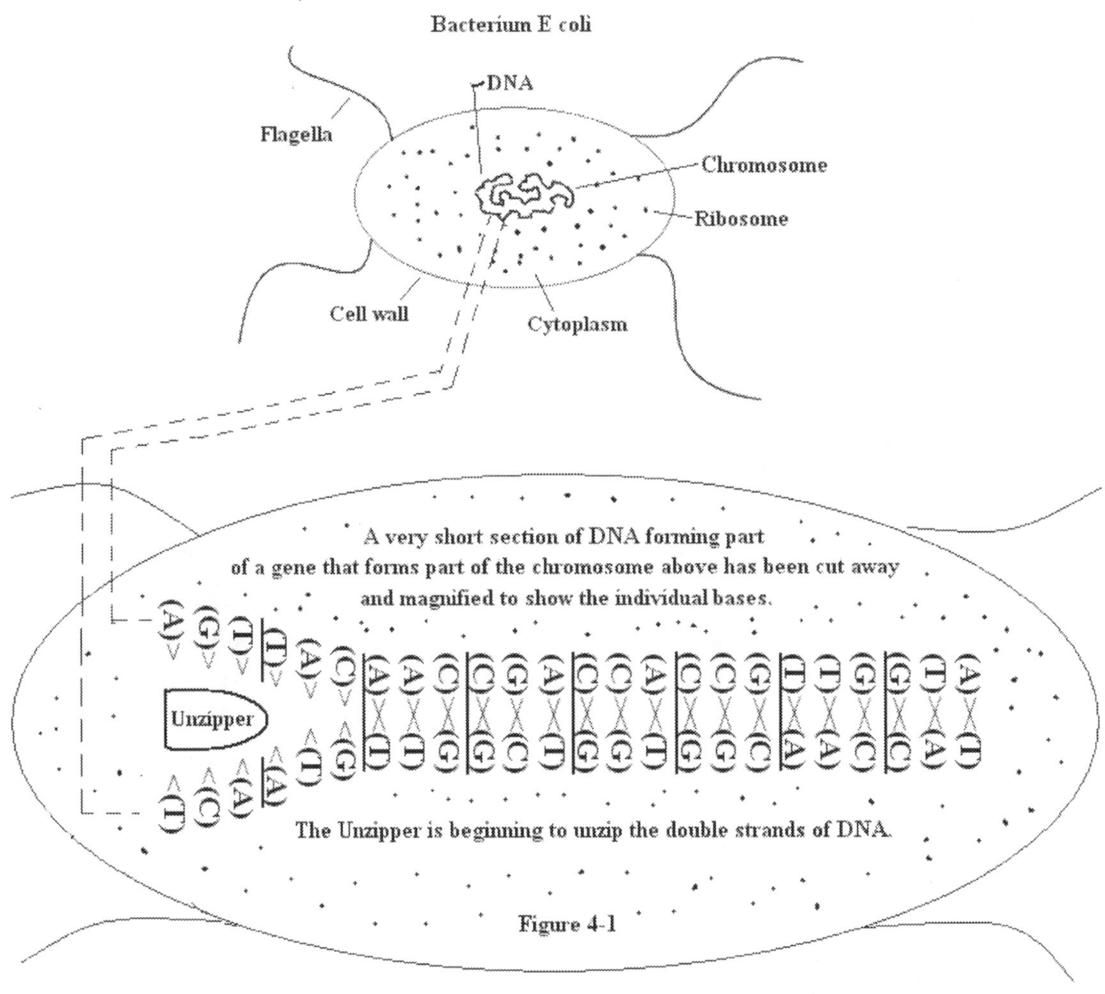

Figure 4-1

The smallest diagram above is an E. coli bacterium with the major parts labeled. Note that the single chromosome is all wadded up, but if untangled would take on a circular shape. Below it, is an enlarged version with a short section containing only 24 base pairs of DNA, that has been cut away from the chromosome above. Note, the unzipper is dividing the double strands.

Below, without enclosing the enlarged diagram of replicating DNA in an imaginary bacterium as was done on the previous page, pretend that it is enclosed. You can see that to the right of the diagram, the original old DNA is continuing to be divided by the unzipper and that two new double DNA strands are being formed on the left. The new double strand on the bottom is called the leading strand and is forming towards the replication fork, as noted by the arrow. The new strand on top, labeled the lagging strand, is being built in sections away from the replication fork (see arrow), with gaps occurring between each section. The lagging strand gets its name because closing the gaps later by the connector slows down the rate of its construction. However, when the gaps are connected end to end, the lagging strand will look exactly like the leading strand.[1]

After the unzipper divides the two strands of the original DNA exposing the "teeth" or bases of each single strand, they act as two separate templates for individual complimentary bases to attach. This process can be visualized in the diagram below, by noticing the complimentary bases approaching the exposed "teeth" or bases that appear in the single strands noted just above and below the unzipper. Since there are about 4,500,000 base pairs in the single circular chromosome of an E. coli bacterium, this means that this number of complimentary bases must be added to each side of the exposed strands, making a total of 9,000.000. Since it has been well documented that E. coli can reproduce itself in 20 minutes or 1200 seconds, this means that 7500 base pairs must be added every second for this feat to occur in the time noted. The roof shaped line at the

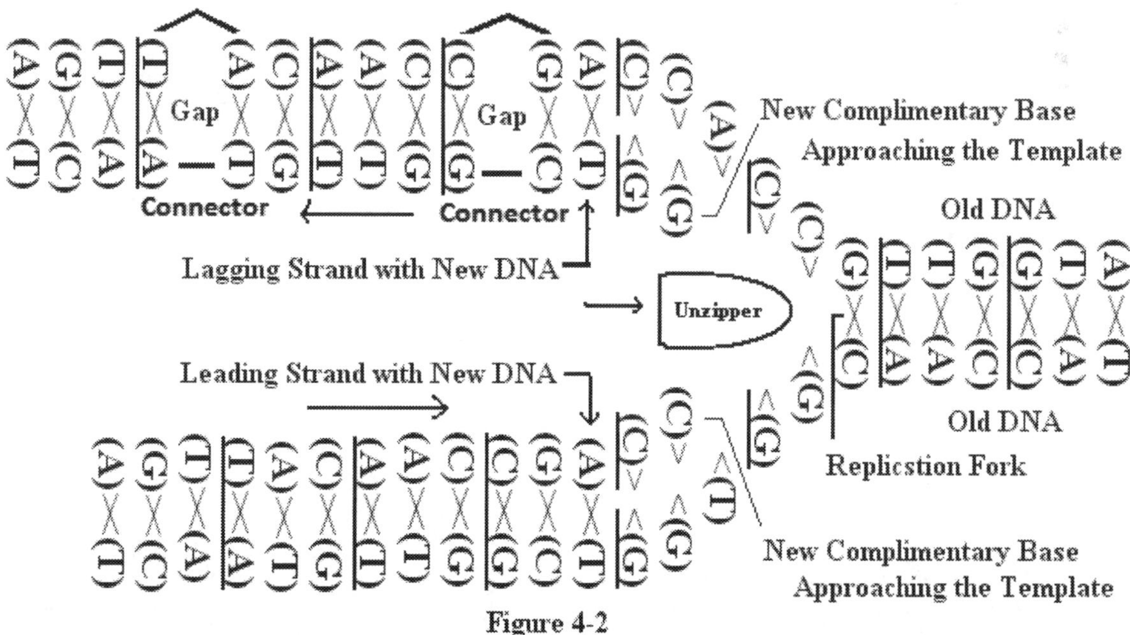

Figure 4-2

top of the diagrams indicate that there is no gap between the base pairs under the roofs . The gaps are in the lagging strand forming below.

Keep on pretending that the next two diagrams are enclosed in a bacterium, See the unzipper has moved almost completely through this short DNA segment. The connector is getting ready to connect the separated segments in the lagging strand. When these processes are finished, two complete genomes or patterns of this prokaryote will exist in the same cell.

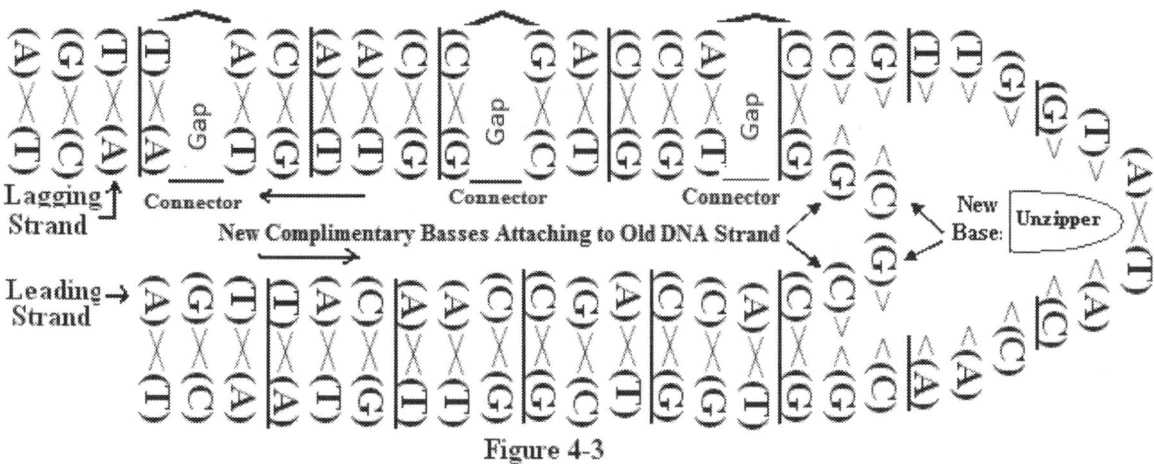

Figure 4-3

Below the unzipper has moved completely through the DNA segment and the connectors have connected the ends of each section together closing the gaps in the lagging stand. Now the cell contains two complete genomes and is now ready to divide in two, with one complete DNA genome going to each daughter cell.

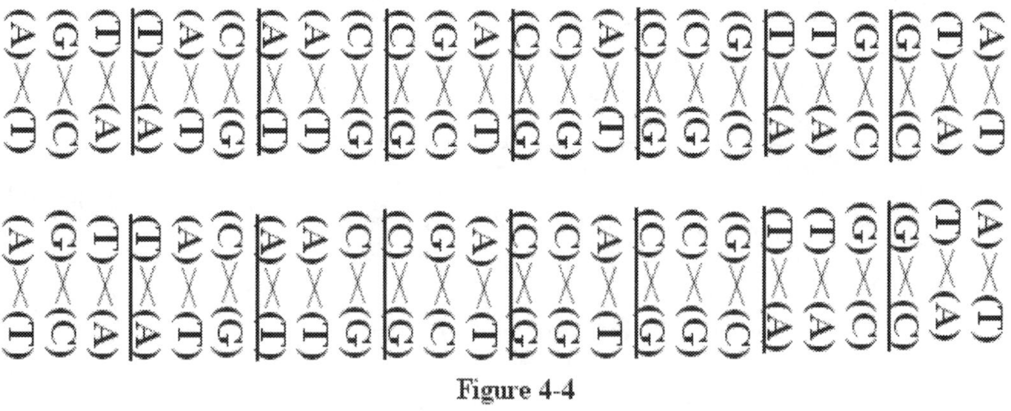

Figure 4-4

The biological processes illustrated above don't just magically happen. Besides the unzipper and the connector, both proteins, there are at least five other protein helpers all with scientific names that pull off this amazing feat. Learning names isn't necessary to see what happens.

The cellular diagram below displays indentations forming at each end of the cell along with a fission line where the division will take place. Note that each half of the dividing cell is getting a complete DNA genome.

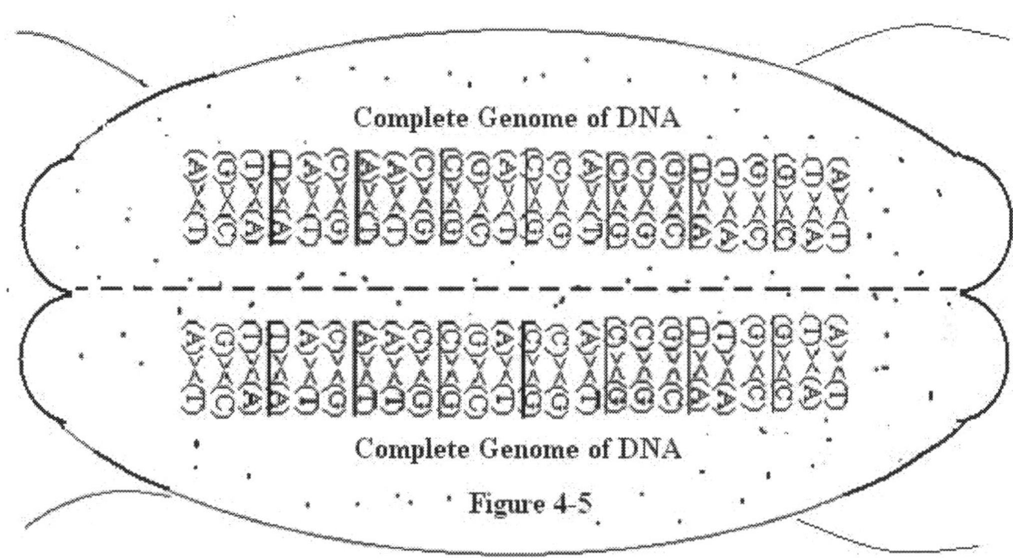

Figure 4-5

Cellular division is complete as noted in the next diagram. Each half got a complete genome.

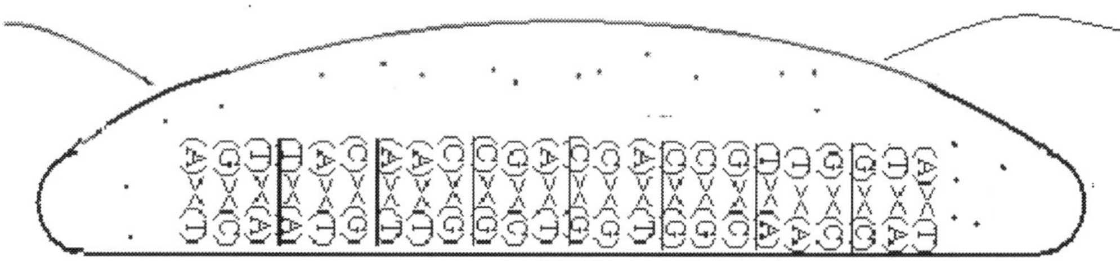

The cell has divided in two with each containing a complete genome.

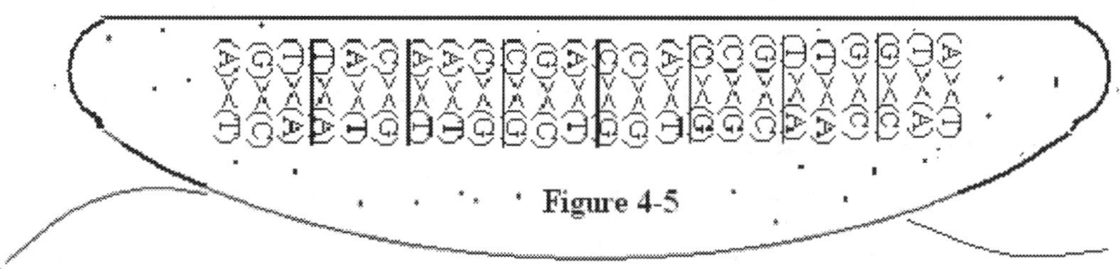

Figure 4-5

Each half will now grow into a mature cell that will be able in turn to divide in two. As previously noted, E coli bacteria, when located in a friendly environment are able to replicate by cellular division every twenty minutes. One bacterium becomes two. Two become four. Four become eight and so on. In a few hours there are actually billions of them.

The terminal ileum (small intestine) and colon (large intestine) in humans and all warm blooded animals is an ideal place for E coli to replicate[2]. A friendly symbiotic relationship is present in this situation. We supply the bacteria with a warm place to live along with plenty of food and they help us digest our food. As fast as they replicate in our guts and "swim" up the fecal stream, they are swept out in our feces where most of them die. If their reproductive cycle was not interrupted in this way by their demise, theoretically one E. coli weighing 10^{-12} grams undergoing exponential reproduction, that is replicating itself every 20 minutes would produce offspring weighing 4000 times the weight of the earth in 48 hours[3].

Nucleotides and Homochirality

Chemistry of the carbon atom is known as organic chemistry. Many organic molecules are three dimensional, forming configurations, which are mirror images of each other, just like your two hands. They are designated respectively, as levorotatory (levo) and dextrorotatory (dextro), or L and D for short. Just as anyone could recognize the difference between a left handed glove and a right handed glove, so recognition of levo or dextro-configured molecules could be identified if they were visible. Each DNA or mRNA molecule is composed of three parts, one sugar (ribose), one phosphate group, and one base. There are five different kinds of bases, namely, adenine (A), thymine (T), guanine (G), cytosine (C), and uracil (U). These can form five different kinds of molecules, each known as a **nucleotide**. All are dextrorotatory in configuration. Any group of molecules with the same configuration are said to exhibit **homochirality**.

When two DNA nucleotides containing bases (A) and (T) are oriented across from each other, the bases can form chemical bonds between themselves making a base pair, A><T, like those noted in chapter 3. The same is true for nucleotides containing (G) and (C). They can form base pair bonds between G><C. When the bases of two nucleotides join together, a double molecule is made and when multiple double molecules attach, one on top of the other, they form the double strands of DNA. RNA polymerase derives the dextrorotatory single strand of mRNA from the templet strand of DNA. The dextrorotatory mRNA nucleotides only react chemically with levorotatory amino acids (except for glycine), and thereby only make levorotatory configured proteins. These examples demonstrate the importance of chirality, which will become more obvious in chapters that follow.

Chapter 5

DNA Replication in Eukaryotes

Mitosis is defined as a process in which the replicated genetic material of a eukaryotic cell is divided forming two separate and equal sets of identical chromosomes.[1] Cytokinesis, or cell division, follows.[2] Mitotic cell division takes place in the life of every eukaryotic cell, be it a "lowly" single-celled eukaryote such as an amoeba or a "sophisticated" brain cell in the human brain. The nucleus, found only in eukaryotes, is the centerpiece of mitosis and is the location of the paired chromosomes. Technically, we cannot term cell division of bacteria (prokaryotes) as mitosis because bacteria have no nuclei and their singular chromosome is not paired. Nevertheless, bacterial cell cycles resemble eukaryotic cell cycles in three basic events:

1. Cell growth
2. Chromosome replication
3. Cell division[3]

The method, frequency, and timing of cell division vary dramatically between various life forms. The bacterial cell (a prokaryote, not a eukaryote), under optimum circumstances, may reproduce itself in cell division as often as every thirty minutes or less.[4] Bacterial cells simply replicate their single circular chromosome and then divide by cytokinesis. However, nerve cells of the human brain apparently cease dividing after the brain has reached maturity. This is why brain damage, from whatever cause, is so devastating and permanent once nerve-cell death has occurred. Total repair is impossible without brain-cell division to replace the damaged cells. On the other hand, the rapidity with which bacteria can multiply by cell division explains how a small cut contaminated by certain kinds of bacteria can become infected so rapidly.

The cell wall, chromosomes, and ribosomes are structures common to prokaryotes and eukaryotes, but only eukaryotes have organelles, one of which is a nucleus, the walls of which temporarily break down during mitosis. Remember that when the parent eukaryotic cell divides, each new daughter cell gets an equal share of each kind of the organelles along with half of the newly replicated chromosomes. Mitosis is about dividing the doubled DNA of one parent eukaryotic cell into two eukaryotic daughter cells. Cytokinesis divides not only the replicated DNA between the two daughter cells, but the cytoplasm containing the organelles as well.[5]

An example of normal eukaryotic cells that reproduce themselves in the process of mitosis is the erythropoietic cell of the bone marrow. They become red blood cells. The average life span of a red blood cell is about 120 days. As the aging red blood cell population breaks down in our spleens like worn-out cars in a junk yard, our bone marrow replaces these worn-out red blood cells with new ones, like brand new cars coming off the assembly line. The average adult male has about five liters of blood in his vascular system. One liter contains one million micro-liters. Five liters of blood equals five million micro-liters. In the average adult male, one micro-liter of blood contains about five million red blood cells. From these approximate numbers, we can calculate that 5 million micro-liters x 5 million = 2.5×10^{13} red blood cells in 5 liters of blood. If we divide 2.5×10^{13} by 120, the average life span of an RBC, we get 2.0833×10^{11}, which is the number of new red blood cells made in one day. There are 86,400 seconds in twenty four hours, so by dividing 2.0833×10^{12} by 86400 we get 2,411,000 or about 2.4 million red blood cells, the number made every second in an average adult male. During each second that an average adult male reads this information, approximately 2,400,000 new red blood cells were manufactured by his bone marrow and about the same number of red blood cells were broken down in his spleen. These figures, of course, differ somewhat depending on a person's sex, size, and age. These numbers are awesome. They may help you appreciate the enormity of this dynamic process and aid you in getting the picture of how important mitosis really is.

In 1879 Walter Flemming first visualized mitosis as he studied the growing cells of salamander larvae. These observations happened in a roundabout manner starting more than 20 years before. At that time a young man named William Perkins was trying to produce quinine synthetically. Quinine, one of the first drugs discovered and used for the treatment of malaria, was obtained from the bark of a tree. In his attempt to produce synthetic quinine, Perkins began mixing various chemicals and discovered when he mixed potassium dichromate with aniline, a purple goo resulted. When this substance was mixed with alcohol, a beautiful purple solution developed which turned out to be an excellent dye, but had no effect on malaria.

Nearly 25 years later, Walter Flemming tried staining cells of salamander larvae with Perkins's dye to see what they would look like under the microscope. Suddenly the nuclei of these cells became visible because they had absorbed the dye. By continuing his microscopic visualization of growing cells in various stages, he was able to visualize nuclear material separating into thread-like divisions that he named chromosomes. Flemming noted that as the chromosomes reproduced themselves, a division occurred so that half of the thread-like chromosomes migrated to one end of the cell and the other half to the opposite end. Following this, the cell divided itself into two separate daughter cells, each containing half of the chromosomes. Flemming called this process mitosis after the Greek word for thread.

Flemming wrote about his findings in 1882. Even though Mendel had done his genetic experiments in the 1860s, his work, though published, was not to be discovered until around 1900. Because of this delay in the promulgation of scientific knowledge, Flemming did not associate his discovery of chromosomes and mitosis as being associated with heredity.

Mendel called the mechanisms that transmitted genetic traits as factors. However, it was Walter Sutton, a biologist from America, who presented evidence that the factors first described by the Austrian monk Mendel, were located on the chromosomes which had first been described by the German cytologist Flemming. Walter Sutton did this work in 1902; however, it was Wilhem Johannsen, a biologist from Denmark, who first called Mendel's factors genes. In 1912 Thomas Hunt Morgan proved Sutton correct. From that time on, biologists were certain that genes located on chromosomes were responsible for individual traits of inheritance.[6]

The process of mitosis has two divisions, **interphase** and the **M phase**. During interphase the DNA is doubled, that is where all of the linear chromosomes are duplicated in a similar fashion as is the case with prokaryotes. The M phase is the mitotic phase and describes how the duplicated chromosomes are pulled apart in preparation for cell division.[7]

DNA is reproduced during interphase[8] by the aid of at least seven helpers.[9] Each one of which is a protein. Unlike the single circular chromosome containing the DNA of the prokaryote, eukaryotic DNA comes in multiple paired linear chromosomes, called homologous chromosomes. However, like the prokaryote, the double stranded DNA of each of the eukaryotic chromosomes must first be split apart forming two single strands. These then act as templates on which two new double strands of complimentary DNA are made. As with prokaryotes, the protein complex that splits or unzips the double strands of DNA, we'll call the "unzipper." It starts at one end of a given linear homologous chromosome and continues splitting the two strands apart as it proceeds through the DNA. As soon as the two single strands are separated, they act as templates on which the other protein helpers go to work placing complimentary bases on each exposed base of the template. Their activity forms multiple base pairs attached to each other, until the whole chromosome is duplicated, which produces two chromosomes exactly like the original.

Below is a diagram taken from prokaryotic replication, which can also show how a very short section of eukaryotic DNA is duplicated. The main differences between the two are that prokaryotic DNA is located in a single circular chromosome whereas the DNA of eukaryotes is stored in a nucleus in multiple linear chromosomes. Otherwise this portion of DNA replication between these two very diverse groups of biota is similar.

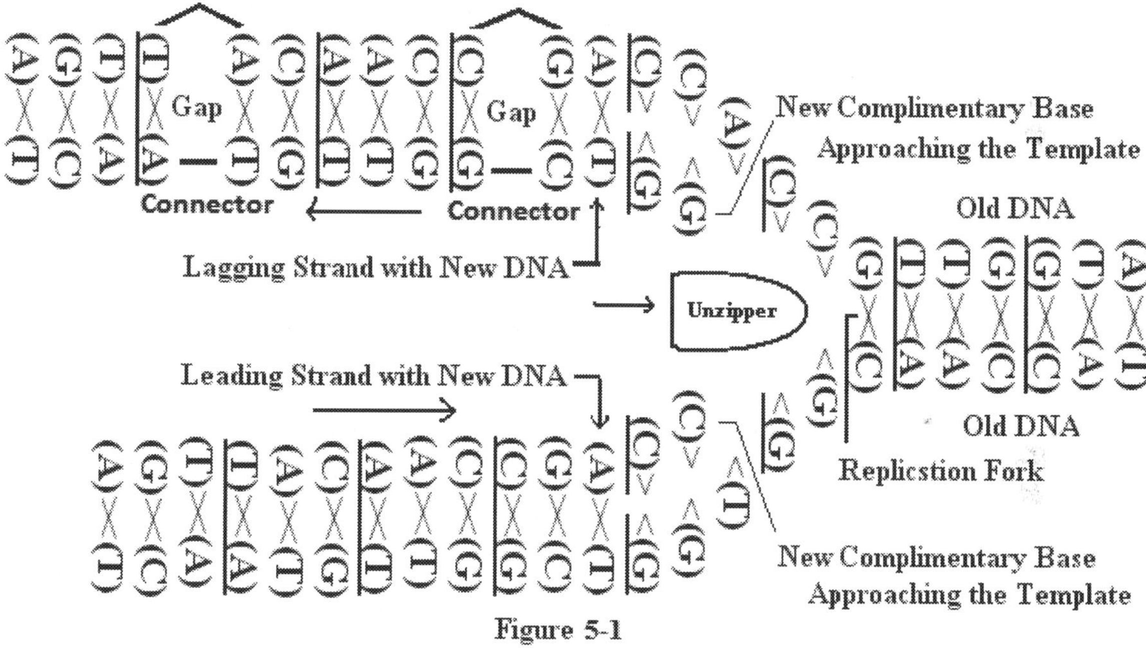

Figure 5-1

The M phase is divided into four other sub-phases, prophase, metaphase, anaphase, and telophase, which will be illustrated later. It is interesting to note that before the advent of chemical analysis of the various mitotic phases, interphase was thought to be a resting phase of mitosis.[10] These wrong conclusions were drawn from observations viewed with a light microscope that could not detect the chemical activity going on during interphase. The activity of the last four phases could be seen with an ordinary microscope and therefore they were thought to be the active phases. Now, we know that interphase is probably the most active phase of all.

Mitosis, or the M phase, actually begins after interphase when all of the homologous paired chromosomes have been duplicated. This M phase involves the separation of the newly formed chromosomes and how they become equally divided in two. This insures that when cytokinesis occurs each new cell will receive half of the doubled DNA and half of the organelles.

The diagram on the next page takes one pair homologous chromosomes through interphase, prophase, metaphase, anaphase, and telophase. Telophase, the last one, happens when the previously disintegrated nucleus is restored in the two newly formed cells. Understand that the two vertical lines at the top of the diagram each represent double strands of DNA, which contain millions of complimentary base pairs. Also understand that the number of linear paired chromosomes varies between different species of eukaryotes, some with as few as one pair and others with many. Human cells contain 23 pairs or a total of 46 chromosomes. Keep in mind that what the succeeding diagram displays as happening in only one pair of homologous chromosomes is occurring simultaneously in all of the chromosomes in that cell at the same time.

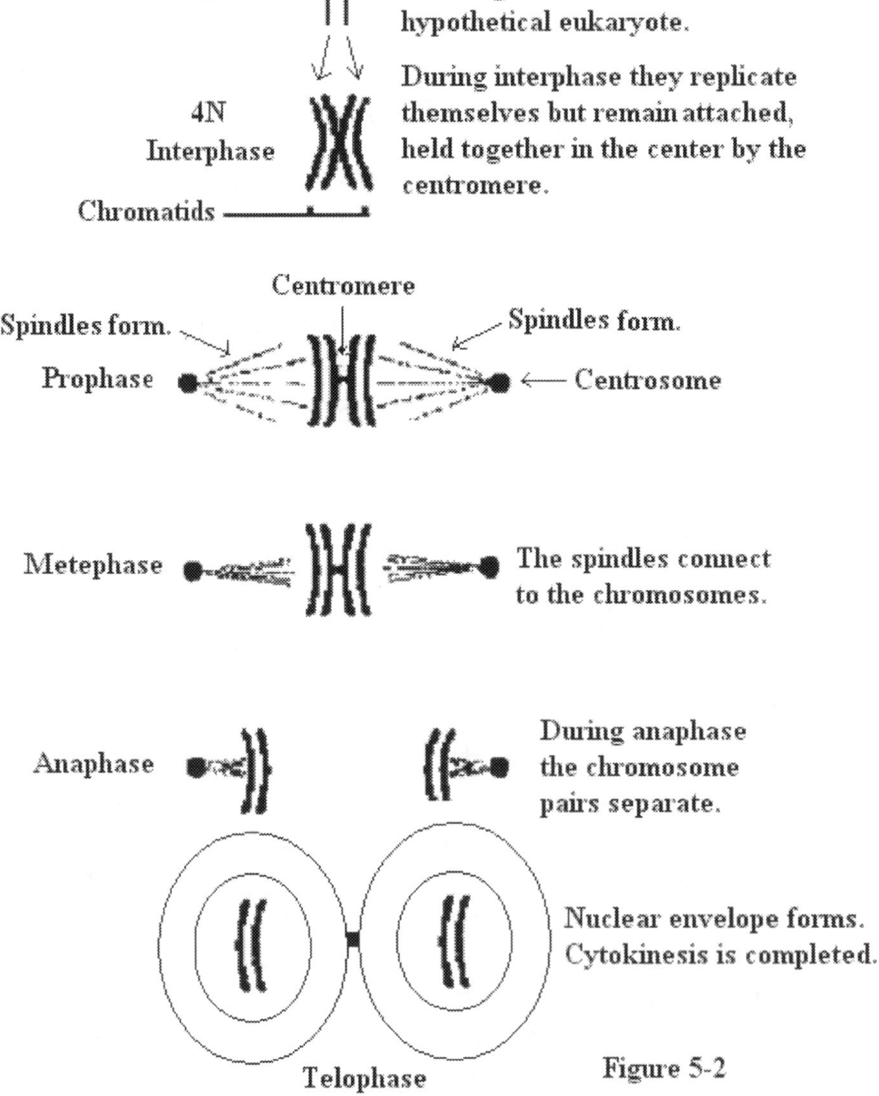
This diagram traces one pair of homologous chromosomes through the process of mitosis, which happens inside a cell. The cell is not shown.
2N
The two lines represent one pair of homologous chromosomes in a hypothetical eukaryote.
4N
Interphase
During interphase they replicate themselves but remain attached, held together in the center by the centromere.
Chromatids
Centromere
Spindles form.
Spindles form.
Prophase
Centrosome
Metephase
The spindles connect to the chromosomes.
Anaphase
During anaphase the chromosome pairs separate.
Nuclear envelope forms. Cytokinesis is completed.
Telophase
Figure 5-2

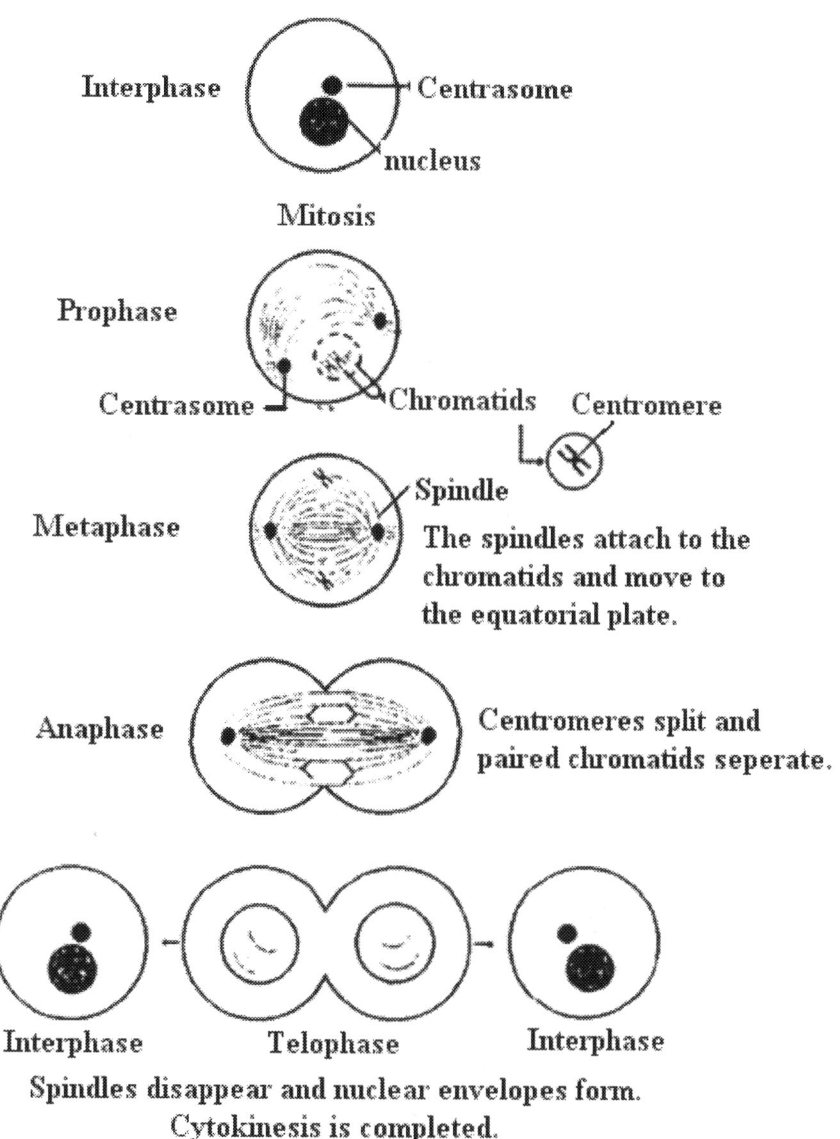

Figure 5-3

The diagram located above displays a pictorial sequence of mitoses starting just after interphase. With interphase complete, prophase, the first sub-phase of the M phase, is ready to start. During prophase, the centrosome divides in two. It is represented as the black dot next to the larger dot, the nucleus, which can be seen in the center of the drawing of the cell at the top of the page. The two centrosomes separate and migrate to opposite poles of the cell. About this time, the duplicated chromosomes condense into clumps, easily visible under the ordinary light microscope. When each pair of homologous chromosomes duplicates, it forms a clump containing four homologous chromosomes (4N). The two pairs remain attached at the centromere like conjoined twins. The paired chromosomes are called chromatids as long as they remain attached at the centromere. Remember that all eukaryotic chromosomes come in pairs. When a given chromosome replicates during interphase, a new pair is formed, and these two pairs of chromatids stay temporarily connected at the centromere. As prophase continues, the nuclear membrane breaks down, leaving the chromatids exposed to cytoplasm. From the centrosomes emanate the asters, composed of microtubules that grow toward the chromatids and become spindles. This completes prophase.

We are now ready to advance to metaphase. Chromatids migrate to the center of the cell and form a circular plate called the spindle equator. Then, the growing microtubules attach themselves to the chromatids, one on each side across from the centromere and begin tugging, trying to pull them apart.[11]

Anaphase starts when the centromere "let's go" and the two paired chromatids split apart. Then the microtubules pull on each one causing them to separate even more.[12] Subsequently, each group of paired chromosomes collect at each of the poles of the cell.

Telophase starts when two sets of chromosomes arrive at the poles and uncoil. The spindles that formed the microtubules then disappear and a new nuclear membrane forms around each of the two groups of paired chromosomes.[13] The cell wall starts to pinch itself into two. When the pinching process of cytokinesis is complete, then telophase is complete and so is mitosis.[14]

This is the same process that makes all the red blood cells (about 2.4 million every second). But remember, there are many millions of developing red blood cells being manufactured inside the bone marrow in a continuous process. However, only about 2.4 million are maturing every second. It takes much longer than a second for a given red blood cell to be made. This is the same process that makes a baby develop in its mother's womb. This is the same process that makes every eukaryote on this planet grow, be it a toadstool, frog, human, or tree.

Summary

1. Mitosis is the process that equally divides the doubled genetic material in cellular division, which takes place in all eukaryotic cells, except sex cells.
2. Eukaryotes all have chromosomes that come in pairs. Humans have 23 pairs, or a total of 46.
3. Mitosis cannot take place in prokaryotes (bacteria) because prokaryotes have only one unpaired circular chromosome and no nucleus.
4. There are two main phases or divisions that occur during replication of eukaryotic cells: Interphase and the M phase. During interphase the cell enlarges and chromosomal (DNA) replication takes place. This process is not visible with a microscope. The M phase or mitotic phase follows, composed of prophase, metaphase, anaphase, and telophase. Then cell division or cytokinesis occurs. This is visible with a microscope.
5. Chromosomal replication is accomplished with the help of seven proteins, the patterns of which also are stored in the DNA.

6. Before any cell can divide into two, the genome containing all of the DNA patterns needed for a given cell must be replicated with extreme accuracy.
7. Eukaryotic cells also have many kinds of microscopic membrane-bound organelles, one of which is a nucleus. At the end of mitosis the eukaryotic cell divides, with each new daughter cell receiving exactly half of the chromosomes and about half of the organelles.
8. When DNA replication starts during interphase the "unzipper," actually a protein, splits the double strand of DNA into two separate strands.
9. Each separated strand acts as a template on which new DNA strands are built, one nucleotide at a time, following the base pair rule. The other 6 proteins carry out the remaining activities of the replication.
10. When the replication fork has moved through the entire DNA, two new DNA molecules have been made. Each has a strand from the old DNA, and each has a new strand built on the template of the old strand, using the base pair rule.
11. Mitotic processes then divide the DNA exactly in half followed by cytokinesis and the forming of two cells with identical genetic material.

Chapter 6

Meiosis

Meiosis is a very special kind of mitosis that sexually reproducing eukaryotes use to produce the female egg and the male sperm; each is called a gamete. This process is called gametogenesis. We have learned that eukaryotic chromosomes come in homologous pairs. Diploid (2N) is the term used to describe the phenomenon that all eukaryotic cells contain paired chromosomes except gametes. Meiosis divides the pairs so each sperm and each egg have only one-half of every homologous chromosomal pair. This is called the haploid number, or 1N. It is contrasted with the diploid number, or 2N, which represents both the homologous chromosomes paired together, which is the normal chromosomal content of non-gamete cells.

Each eukaryotic egg and each eukaryotic sperm only contains half the number of homologous chromosomes or 1N. They unite at fertilization, and a new individual resulting from this union obtains one-half of each homologous chromosomal pair from the female parent and one-half of each homologous chromosomal pair from the male parent. This brings the chromosomal number from haploid, or 1N, to diploid, or 2N. Keep in mind that meiosis always precedes fertilization and usually takes place in both the male testicles and female ovaries as the gametes are being made. Meiosis is the process that divides the homologous chromosomal pairs in half in the egg and sperm. However this is only part of the story.

Meiosis also mixes up the genes between the homologous chromosomal pairs in a random way which causes variations of inheritance.[1] This mixing takes place in the primordial sex cells before they become haploid or 1N. After this mixing occurs, the mixed up chromosomes split apart producing haploid or 1N individual unpaired chromosomes, each containing a mixture of genetic material. This process of mixing genetic material during meiosis is called crossing over. However, both fertilization and meiosis contribute to the mixing of genetic material. Fertilization mixes eukaryotic material because half is contributed by the male parent and half by the female parent. The crossing-over process in the production of the two kinds of gametes, plus the mixing of genetic material caused by fertilization, triple mixes the genetic material with each generation. This makes all individual eukaryotic organisms that sexually reproduce to have different genetic compositions, except for identical twins.

One of the first research scientists to microscopically demonstrate the crossing-over process of meiosis and understand its significance was Barbara McClintock. Born June 16, 1902, in Hartford,

Connecticut, McClintock was the youngest of three sisters. As a young girl she was active in sports and music. Later her family moved to Brooklyn, New York, where she attended Erasmus Hall High School. It was here that she discovered her love of science. After graduating in 1918, Barbara seriously studied at Cornell University as a botany major. The study of plants was to be her life work.[2]

Even though a female in a male dominated environment, she pressed on obtaining a Bachelor of Science degree in 1923 and a Masters degree in 1925. McClintock obtained her Ph.D. degree in 1927, in the field of plant genetics.[3] With a doctorate in botany by age 24, McClintock used maize as her source of research material. She raised her own corn, studying the microscopic configuration of the 10 chromosomes (haploid) of this eukaryotic plant.[4, 5]

Although Thomas Morgan demonstrated the rearrangement of chromosomes known as crossing over in 1917,[6] it was McClintock, along with her associate Harriet Creighton, who described the exchange of genetic information between homologous chromosomes that occurs during meiosis. The location where two chromosomes cross-over and exchange genetic information is called the chiasmata (Janssens, 1909).[7] In 1931, McClintock and Creighton had their findings published in the Proceedings of the National Academy of Science.[8]

In the description of gametogenesis that follows, an imaginary eukaryote with only one pair of homologous chromosomes will be used as an example. Actually, most eukaryotes have multiple pairs of homologous chromosomes. However, what is described as occurring in one pair applies to all the rest. During the first phase of meiosis, each of these two individual homologous chromosomes reproduces itself, making four. However, each of the duplicated chromosomes stays connected to its predecessor like two sets of conjoined twins. Each duplicated chromosome that stays connected to its twin at the center is called a chromatid and each of these two sets of conjoined twins stays close together, making what is called a tetrad.

Next, in this conjoined twin analogy, equivalent segments of one set of conjoined twins exchange places with equivalent segments of the other set of conjoined twins. Remember, there are actually four chromatids (chromosomes) involved. Because one set of these analogous conjoined twins originated from the mother and the other set originated from the father of the individual in whose gonad this gametogenesis is occurring, the exchange produces a random mixture of genetic material between the two sets of analogous conjoined twins. In other words, an equivalent segment (a gene or genes) of one twin in one set trades places with an equivalent segment (a gene or genes) in another twin in the other set. Equal amounts of genetic material, located at an identical place on one twin, will only trade with an equal amount of genetic material at the identical location on the other twin. The chromatids composing the tetrad now separate in pairs from the tetrad. This primordial sex cell next divides with each new cell receiving a pair of chromatids. Then in the male gonad, the two chromatids separate at their attachments, and each of the two cells divides again making four germ cells, each containing one chromosome or 1N. Each of these four cells contains one chromosome with mixed genetic material, some of which came from the father and some from the mother of the individual in which gametogenesis is taking place. To repeat, this process of mixing genetic material in meiosis is called crossing-over. Gametogenesis in the female differs slightly.

In true meiosis, no new genetic material is added; it is only mixed. Two cell divisions that occur in sequence produce this change. The primordial sex cell is 2N, and each resulting gamete is 1N. Each of these divisions has a prophase, metaphase, anaphase, and telophase, similar to mitosis. Meiosis in the female differs only slightly from the male, so when you understand meiosis in the male, it will be easy to change over to meiosis in the female.[9] Figure 6-1 displays a series of diagrams. These will help you understand how meiosis in the male accomplishes the dual task of mixing up the genes between each of the two paired chromosomes followed by making four haploid (1N)gametes from one original diploid (2N) cell. In Figure 6-1, each circle represents a cell with a diagram of one pair of eukaryotic chromosomes in various stages of meiosis. These diagrams show a hypothetical eukaryotic cell with only one pair of chromosomes. What is shown in this pair of chromosomes is occurring simultaneously with all the other chromosomes during meiosis in a given species, depending upon the number of chromosomes present. Simply start at the top of the diagrams (Figure 6-1) and follow the instructions to see how the genetic material is mixed in the case of male gametes. You also can see how each gamete becomes 1N. Figure 6-2 illustrates gametogenesis in the female. In the male, four gametes, all 1N, are made from one original cell. In the female, only one gamete with mixed genetic material develops because two small polar bodies are expelled. This is done to conserve cytoplasm. However, the expulsion of polar bodies does subtract some genetic material from the genetic pool. Because of this during female gametogenesis, a defective gene has a 75% chance of being expelled in one of the polar bodies.

During fertilization, the male sperm supplies one-half of each chromosomal pair, and the female egg supplies the other half. This raises the haploid number 1N in each gamete to the diploid number 2N in the fertilized egg, called a zygote. Each species of eukaryote has a number of paired chromosomes specific to that species. When fertilization has taken place, all the information to build a new individual eukaryote is present in the zygote. As long as the zygote is in a place where it can grow by the process of mitosis, it will mature into an individual consistent with the species of eukaryote from which its chromosomal number is derived. These basic principles are true for just about every species of higher eukaryote. The males in bees, wasps, and ants are the exceptions. In these males, each of their body cells is only haploid. This only shows that the DNA sequence is more important than the total chromosomal number.

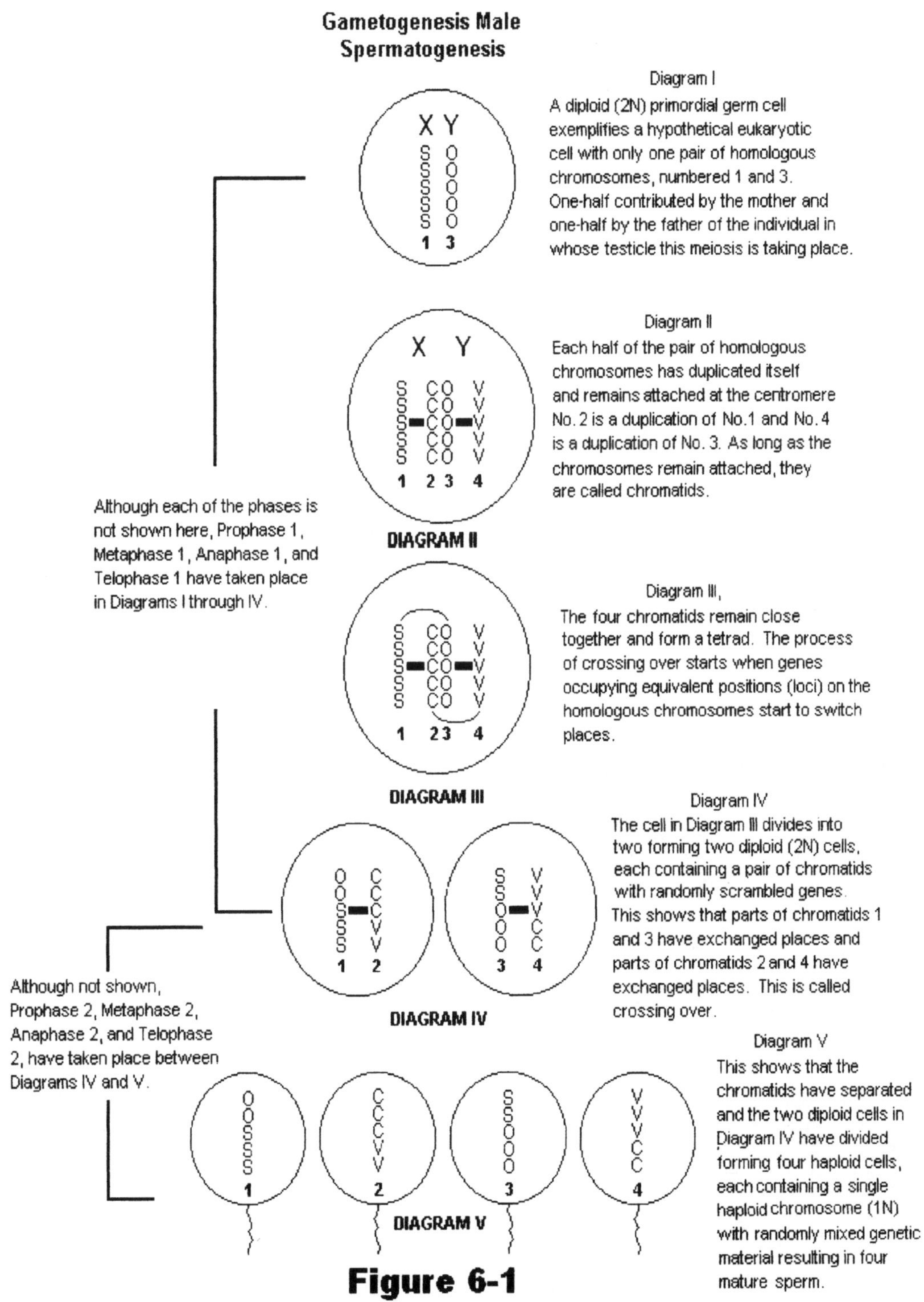
Gametogenesis Male
Spermatogenesis
Diagram I
A diploid (2N) primordial germ cell exemplifies a hypothetical eukaryotic cell with only one pair of homologous chromosomes, numbered 1 and 3. One-half contributed by the mother and one-half by the father of the individual in whose testicle this meiosis is taking place.
Diagram II
Each half of the pair of homologous chromosomes has duplicated itself and remains attached at the centromere No. 2 is a duplication of No.1 and No.4 is a duplication of No. 3. As long as the chromosomes remain attached, they are called chromatids.
DIAGRAM II
Although each of the phases is not shown here, Prophase 1, Metaphase 1, Anaphase 1, and Telophase 1 have taken place in Diagrams I through IV.
Diagram III,
The four chromatids remain close together and form a tetrad. The process of crossing over starts when genes occupying equivalent positions (loci) on the homologous chromosomes start to switch places.
DIAGRAM III
Diagram IV
The cell in Diagram III divides into two forming two diploid (2N) cells, each containing a pair of chromatids with randomly scrambled genes. This shows that parts of chromatids 1 and 3 have exchanged places and parts of chromatids 2 and 4 have exchanged places. This is called crossing over.
DIAGRAM IV
Although not shown, Prophase 2, Metaphase 2, Anaphase 2, and Telophase 2, have taken place between Diagrams IV and V.
Diagram V
This shows that the chromatids have separated and the two diploid cells in Diagram IV have divided forming four haploid cells, each containing a single haploid chromosome (1N) with randomly mixed genetic material resulting in four mature sperm.
DIAGRAM V

Figure 6-1

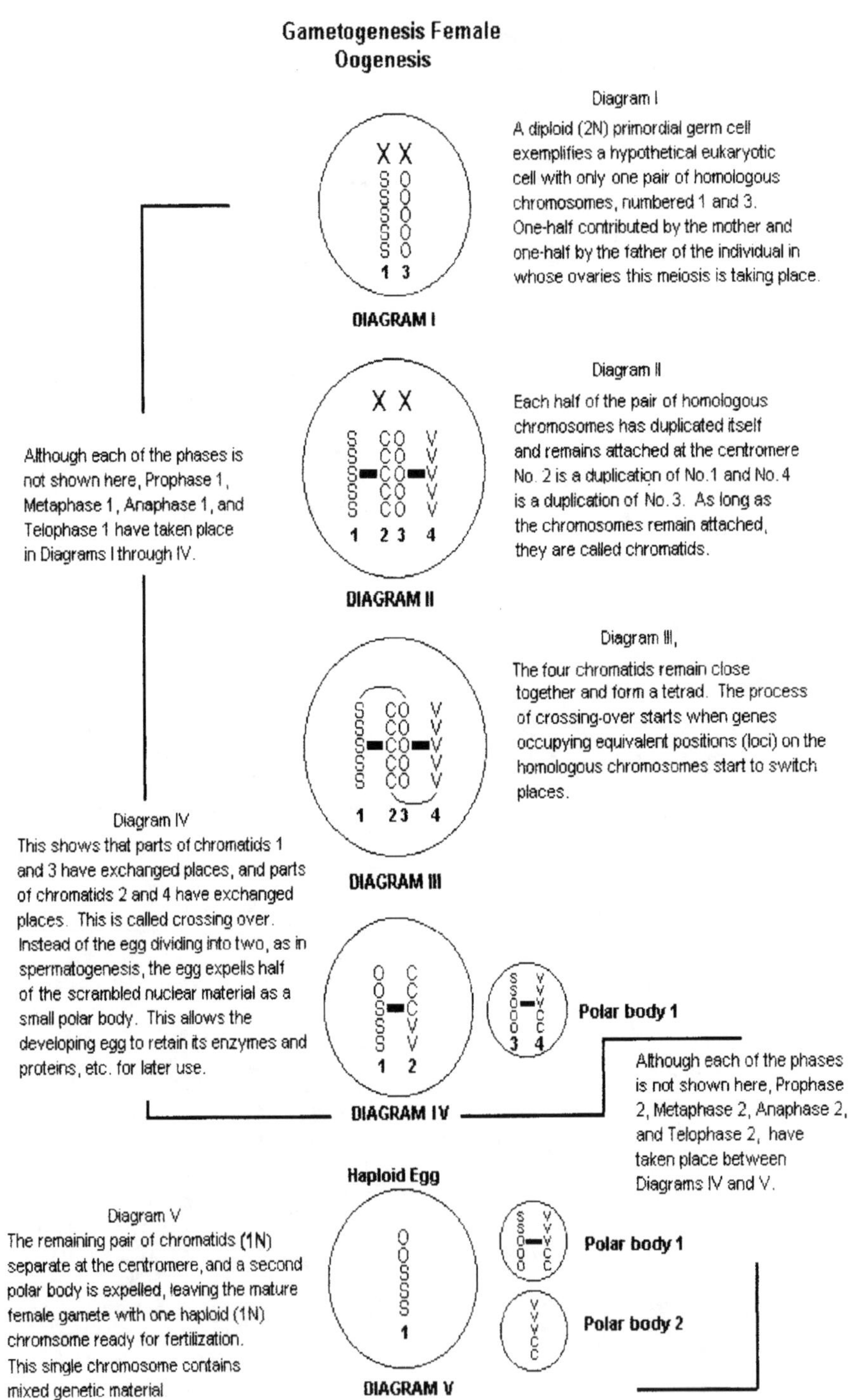
Gametogenesis Female
Oogenesis
Diagram I
A diploid (2N) primordial germ cell exemplifies a hypothetical eukaryotic cell with only one pair of homologous chromosomes, numbered 1 and 3. One-half contributed by the mother and one-half by the father of the individual in whose ovaries this meiosis is taking place.
DIAGRAM I
Diagram II
Each half of the pair of homologous chromosomes has duplicated itself and remains attached at the centromere No. 2 is a duplication of No.1 and No.4 is a duplication of No.3. As long as the chromosomes remain attached, they are called chromatids.
Although each of the phases is not shown here, Prophase 1, Metaphase 1, Anaphase 1, and Telophase 1 have taken place in Diagrams I through IV.
DIAGRAM II
Diagram III,
The four chromatids remain close together and form a tetrad. The process of crossing-over starts when genes occupying equivalent positions (loci) on the homologous chromosomes start to switch places.
DIAGRAM III
Diagram IV
This shows that parts of chromatids 1 and 3 have exchanged places, and parts of chromatids 2 and 4 have exchanged places. This is called crossing over. Instead of the egg dividing into two, as in spermatogenesis, the egg expells half of the scrambled nuclear material as a small polar body. This allows the developing egg to retain its enzymes and proteins, etc. for later use.
Polar body 1
Although each of the phases is not shown here, Prophase 2, Metaphase 2, Anaphase 2, and Telophase 2, have taken place between Diagrams IV and V.
DIAGRAM IV
Haploid Egg
Diagram V
The remaining pair of chromatids (1N) separate at the centromere, and a second polar body is expelled, leaving the mature female gamete with one haploid (1N) chromsome ready for fertilization. This single chromosome contains mixed genetic material
Polar body 1
Polar body 2
DIAGRAM V

Figure 6-2

The genetic differences between the two sexes are explained next. Every cell containing a nucleus in females contains a double "XX" pair of chromosomes, and every nucleated cell in males contains an "XY" pair of chromosomes. This is the genetic difference between the two sexes. Sperm, being haploid, contain either an "X" or a "Y" chromosome. Haploid eggs always contain only "X" chromosomes. If a sperm containing an "X" chromosome fertilizes the egg, the resulting individual will always be female. If a sperm containing a "Y" chromosome fertilizes an egg, the resulting individual will always be male. From this we can see that it is the father's sperm that determines the sex of the next generation. However, besides the exception already noted above with respect to bees, wasps, and ants there is the very surprising exception found in birds. The cells of all female birds contain XY chromosomes and the cells of all male birds contain XX chromosomes. So as a youngster, when you were informed about the birds and the bees, you were given much misinformation. In the case of birds, it is the female bird's egg, and not the male bird's sperm, that determines the sex of the next generation.[10]

Non-identical twins in humans occur when two eggs ripen in the ovary at the same time and both become fertilized. Usually, in humans, only one egg ripens with each menstrual cycle. But sometimes two or more eggs develop simultaneously. If these eggs get fertilized, each by a different sperm, the genetic material in each egg will be different. Also, depending upon whether or not a sperm containing a "Y" chromosome or an "X" chromosome fertilizes the individual eggs, the sex of the individual twins may vary. Twins developing from this situation are not identical because they have resulted from two separate eggs, each of which has experienced a different meiotic crossing-over formation and each having been fertilized by a separate sperm. Non-identical twins may be either the same sex or different.

Identical twins result when a single egg, fertilized by a single sperm, divides itself into two separate individuals at the first cell division. Identical twins that result from this have the identical genetic make up and are always the same sex.

Meiosis and sexual reproduction triple mix the genetic material without adding any. (However, some genetic material is discarded during gametogenesis of the female egg when polar bodies are expelled). The first mix takes place during crossing over in the formation of each sperm. Crossing over causes the second mix to take place in the formation of each egg. The third mix occurs when a sperm containing genetically mixed homologous chromosomes (1N) unites at the time of fertilization with an egg containing genetically mixed homologous chromosomes (1N). This forms a fertilized egg known as a zygote. This triple-mixing process explains the many differences between one generation and the next.

SUMMARY

1. Gametogenesis is the mechanism by which eukaryotic reproductive germ (sex) cells are generated from primordial sex cells.
2. Meiosis is a part of gametogenesis and is the process by which the various homologous genes are mixed and haploid (1N) germ cells are made.

3. At the beginning of meiosis, each primordial sex cell has paired homologous chromosomes, just like any other eukaryotic cell.
4. Any eukaryotic cell containing paired homologous chromosomes is said to be diploid, or 2N.
5. As meiosis starts in a given primordial sex cell, each individual homologous chromosome in each pair duplicates itself resulting in a 4N cell called a tetrad.
6. These newly duplicated chromosomes remain attached at some point between the two, like a pair of conjoined twins. As long as these two chromosomes remain attached, each is known as a chromatid, instead of a chromosome.
7. At this stage of meiosis, each individual homologous chromosome that formed a pair at the outset is now represented by a pair of chromatids. This means that there are four chromatids (like two pairs of conjoined twins), one pair of chromatids for each homologous chromosome noted at the outset.
8. Next, genes in one pair of chromatids trade places with genes in the other pair of homologous chromatids. This process of exchanging genes is called crossing over.
9. After crossing over is completed, the paired chromatids forming the tetrad separate again, each pair containing the exchanged genes.
10. Next, this primordial sex cell divides. One pair of chromatids goes to one daughter cell and the other pair goes to the other daughter cell. This yields two cells, each diploid, or 2N. Each cell contains chromatids with mixed genes.
11. Next, each of the chromatids splits apart where they had been attached, and each of these resulting cells divides again with only one homologous chromosome from each pair of chromatids going to each cell. Each of these cells is now a mature sex cell that contains a single homologous chromosome instead of a pair. This is known as haploid, or 1N. The process just described is how spermatocytes are made in the male (Figure 6-1). A similar process occurs in the female except that two polar bodies are expelled for each mature egg that is made (Figure 6-2).
12. At fertilization, one haploid (1N) sperm from the male fuses with one haploid (1N) egg from the female, resulting in the zygote which is diploid, or 2N. This means that one-half of each chromosomal pair was supplied by the male and one-half by the female. However, it must be pointed out that in the case of both the male sperm and the female egg, part of each haploid chromosome came from the parents of the male or female who made the respective sex cell.
13. During meiosis no genetic material is added to any chromosome. Crossing over and fertilization simply triple mixes the genetic material in a random way in each zygote with each generation. However, some genetic material is lost when the polar bodies are expelled during female gametogenesis.

Chapter 7

Mutations

You already know that each triplet of nucleotides on the template strand or anti-sense strand of the DNA forms a codon. In actuality, there are no physical divisions in the DNA molecule, but rather two continuous chains of nucleotides forming a spiral shaped double helix. When transcribing a gene from the template strand or anti-sense strand of DNA, RNA polymerase incorporates each successive complementary nucleotide, one at a time, into a single strand of elongating mRNA. Then, a ribosome attaches itself to the mRNA strand and translates it into a protein, not one nucleotide at a time, but by reading each successive triplet of nucleotides. Each of these triplets is a codon that specifies one amino acid. Every biological protein is a specific sequence of amino acids connected end to end as determined by the pattern in the DNA specific to each species. So remember, RNA polymerase transcribes DNA into mRNA, one nucleotide at a time, as contrasted with a ribosome that translates mRNA into amino acids, one codon at a time. Each successive triplet of nucleotides in mRNA forms one codon that is translated into a specific amino acid. The ribosome produces proteins which are long chains of amino acids by linking each successive translated amino acid to the preceding amino acid of the growing chain. A bacterium, such as E. coli, has about 4,300 genes that specify several thousand proteins, or polypeptide chains, used by it in its metabolism.[1] Man has approximately 25,000 genes that are transcribed into mRNA, which in turn is translated into polypeptide chains that make up the many proteins found in the human body.[2]

Any mutation is simply a random mistake made in the DNA, which causes a change to arise.[3, 4] By changes we mean that the mutated DNA of a given individual cell differs from that of its predecessor. There are two basic causes for these changes to occur: external and internal. External causes are basically of three types: radiation, chemical, or viral. [5, 6] These external forces can cause changes to occur in the DNA, at any time. Internal causes result simply when mistakes are made during DNA replication. These same types of transcription or translation mistakes can happen in prokaryotes and eukaryotes. Mistakes in DNA replication can also occur during meiosis in eukaryotes that reproduce sexually. Consider all of these mistakes as accidental and random. Mistakes made in each of these examples are called mutations.

There are three main reasons a mutation of any type is rare. The first is the accuracy of selection of the complementary base pair mechanisms. Second is the ability of DNA polymerase I to proofread. Third is the existence of other DNA repair mechanisms.[7]

The simplest spontaneous mutations are single base pair mistakes made in DNA replications. [8] This kind of mutation simply substitutes one complete complimentary base pair for another in the DNA. It only means that one wrong base pair is substituted for the correct base pair. Estimations are that one successful base pair mutation occurs once in every billion base pair replications. A successful mutation is one that goes undetected by various built-in correcting mechanisms already mentioned and more to be discussed later in this chapter. However, many are unsuccessful because they are corrected, netting about one successful base pair mutation per billion replications. A base pair mutation (also known as a point mutation) would be analogous to a musician playing one wrong note at a given place in the musical composition. Substituting one wrong note for one correct note would not change the total number of notes in the piece of music being played. Likewise, a point mutation does not change the number of base pairs in a given gene; it only changes one codon to another. Therefore, the resulting protein may have only one wrong amino acid in its makeup. This type of mutation can be subtle and may cause no change, minimal change, or severe change.

To understand how a point mutation can cause no change, minimal change, or severe change, take a minute to look at line five in the column labeled Roman numeral II at the bottom of the decoder (Chapter 3, table 3-2). You will see that there are four different codons for proline. You will notice that the DNA anti-sense strand for the first codon for proline is GGA. Now suppose that the point mutation substituted a "G" for the "A" in the codon. The mutated codon "GGG," would still specify proline, so the mutation would not produce any change in the resulting protein. This kind of point mutation is called a silent point mutation.

From here, it does not take much imagination to figure out that some point mutations could be between silent and severe. If a point mutation changes the codon to code for another amino acid that functions similar to the amino acid that the mutation replaced, the resulting protein will continue to perform the same as the non-mutated protein and no ill effects will be noted by the organism. This type of mutation is called a neutral mutation.

If a point mutation changes an amino acid codon in the DNA to one that causes the resulting protein to function poorly, the organism containing it, will suffer untoward reactions or even premature death as a result. This type of point mutation is called a missense mutation. A good example of this type of mutation in humans is the one that causes sickle cell anemia. It causes a significant disruption in a person's life but does not kill that person before birth like a lethal mutation would do. This mutation is most common in people of African descent. There are four polypeptide chains making up the giant life-sustaining hemoglobin molecule: two alpha chains and two beta chains. This mutation involves a change at the sixth position on both of the beta chains, valine substitutes for glutamic acid. [19] Look at the anti-sense strand column of your decoder (Chapter 3, table 3-2), for glutamic acid, Roman numeral III, number fifteen and valine in Roman numeral I, number fifteen. To change the codon for glutamic acid, which is "CTT," into the codon for valine, which is "CAT," the middle "T" in "CTT" only has to be substituted with an. "A" This seemingly inconsequential change, plagues the person with the mutation all the days of his or her shortened life. These patients are always anemic, because the abnormal hemoglobin folds on itself causing the red blood cells to form abnormal shapes, hence the name sickle cell anemia. The spleen destroys these abnormally shaped red blood cells, causing the anemia. Sometimes these cells form clots in small blood vessels, which cause painful episodes.

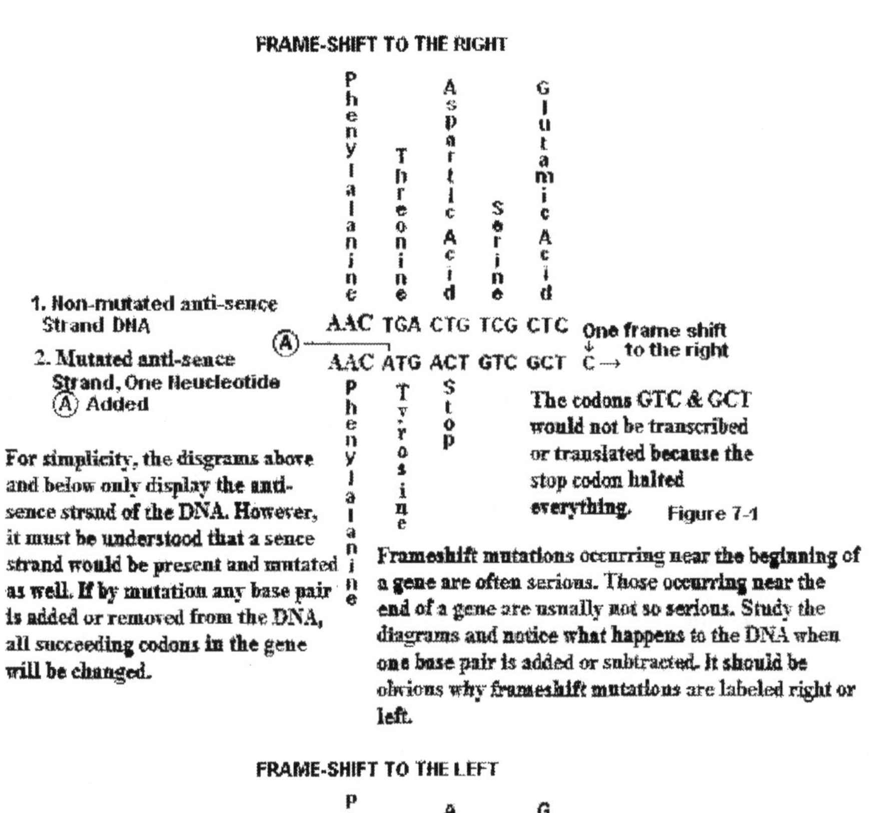

FRAME-SHIFT TO THE RIGHT
Phenylalanine
Threonine
Aspartic Acid
Serine
Glutamic Acid
1. Non-mutated anti-sence Strand DNA
AAC TGA CTG TCG CTC
One frame shift to the right
A
2. Mutated anti-sence Strand, One Neucleotide (A) Added
AAC ATG ACT GTC GCT C
Phenylalanine
Tyrosine
Stop
The codons GTC & GCT would not be transcribed or translated because the stop codon halted everything.
For simplicity, the disgrams above and below only display the anti-sence strand of the DNA. However, it must be understood that a sence strand would be present and mutated as well. If by mutation any base pair is added or removed from the DNA, all succeeding codons in the gene will be changed.
Frameshift mutations occurring near the beginning of a gene are often serious. Those occurring near the end of a gene are usually not so serious. Study the diagrams and notice what happens to the DNA when one base pair is added or subtracted. It should be obvious why frameshift mutations are labeled right or left.

Figure 7-1

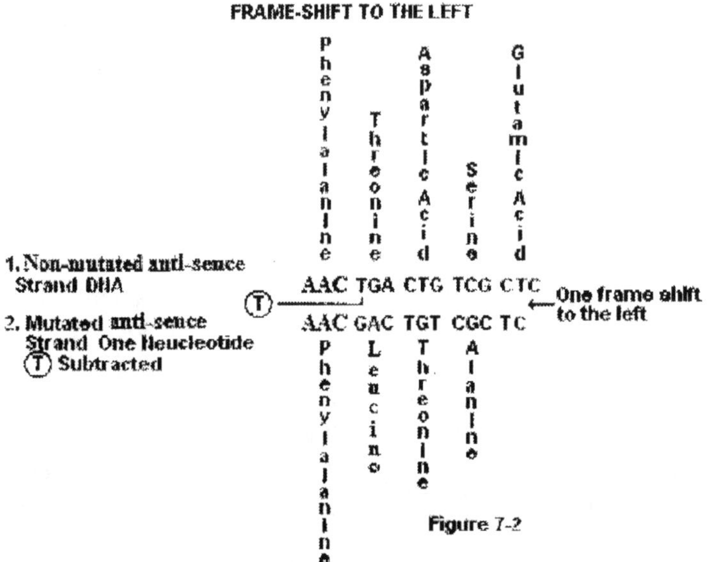

Figure 7-2

Suppose a point mutation changed the codon for cysteine from "ACA" to "ACT." The ACT is a stop codon. What would happen? The mRNA translation would come to a halt, whether the mutation was at the beginning, middle, or end of a gene. Of course in most instances this would be devastating. This type of point mutation is called a nonsense mutation.

The frame-shift mutation also involves one base pair (Figures 7-1, 2). Each individual pair of nucleotides in DNA, or each individual nucleotide in RNA, is considered as a frame. In frame-shift mutations in DNA, one base pair is added or one base pair is subtracted, anywhere in the long DNA molecule. In RNA, a single nucleotide is involved, one being added or one being deleted. This would be like one extra note being added to, or one note being subtracted from a musical score. A frame-shift to the right is where a new base pair is added anywhere in a DNA or one nucleotide is added to the mRNA molecule. Subtract any base pair from the DNA molecule and it will cause a frame-shift to the left. The same rule applies to frameshift mutations in mRNA. Unless a frame-shift mutation occurs near the end of a gene, it will be devastating because it changes all the codons downstream from that point.

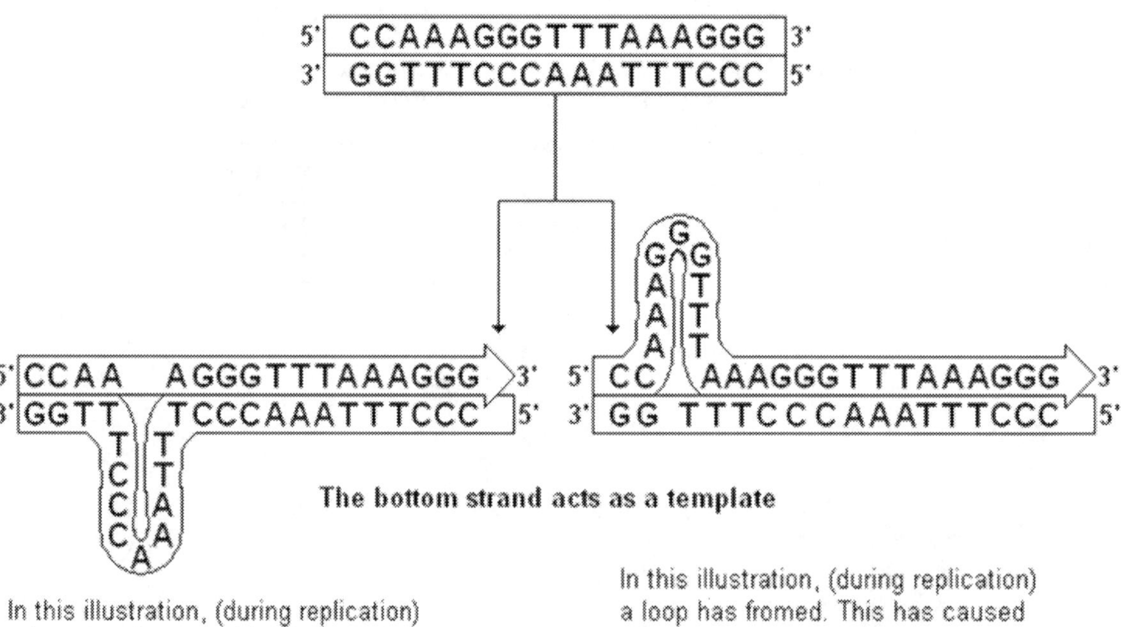
Two Possibe Stem and Loop Mutations
5' CCAAAGGGTTTAAAGGG 3'
3' GGTTTCCCAAATTTCCC 5'
5' CCAA AGGGTTTAAAGGG 3'
3' GGTT TCCCAAATTTCCC 5'
TCCCA
TTAA
5' CC AAAGGGTTTAAAGGG 3'
3' GG TTTCCCAAATTTCCC 5'
GAAAA
GGTTT
The bottom strand acts as a template
In this illustration, (during replication) a loop has formed which has removed nine bases from the new strand.
In this illustration, (during replication) a loop has fromed. This has caused nine bases to be added to the new strand.
This mutation can occur when complimentary regions are present in a given DNA strand.

Figure 7-3

The reason for this is that the ribosome, which translates RNA into protein, reads one triplet of nucleotides at a time, not one nucleotide at a time. When one base pair is added or subtracted from the DNA molecule or one nucleotide is added or subtracted from the mRNA, the ribosome keeps on reading each succeeding three nucleotides as a single triplet, even though at the point of frame-shift mutation, the next set of triplets is partially made from the previous set and part from the succeeding set. This applies to every remaining codon in that gene as translation proceeds. Because prokaryotes have only one circular chromosome, which has no end, a frame-shift mutation in a given gene will be devastating as it affects every succeeding triplet codon remaining in the entire gene. It must be remembered that for a frame-shift mutation to go undetected by the proofreading mechanisms, described later in this chapter, the added or subtracted nucleotide in the template strand or anti-sense strand must be balanced by a corresponding complementary base in the non-template strand or sense strand. In other words, when a frame-shift mutation adds or subtracts a nucleotide from the template strand or anti-sense strand, a corresponding base must be added or subtracted from the non-template strand or sense strand. Frame-shift mutations make the number of nucleotides in the mutated DNA molecule no longer divisible by three[10].

Another kind of spontaneous mutation that can occur during DNA replication is caused by stem and loop structures (Figure 7-3). In each case, a loop forms at a position in either the template strand or replicating strand. If the loop is in the template strand, this can cause the loss of genetic material later. If the loop is in the replicating strand, this can cause addition of genetic material.[11] In our music analogy, this mutation is similar to subtracting or adding several bars of music at a given place in the score. This mutation would not be very likely to add base pairs to the genome, as about 50% of the time base pairs would be added and about 50% of the time subtracted.

Besides the point mutation, the frame-shift mutation, and the stem and loop mutation, there is another called the transpondable element mutation. Transpondable element mutations come in two different categories--insertion sequences and transposons. These types can happen in either prokaryotes or eukaryotes. This is where a protein called transpondase removes a large segment of DNA from one place and reinserts it back into another site in the DNA. Insertion sequences involve only a few thousand base pairs. However, transposons are larger involving multiple genes. Sometimes in eukaryotes this large segment of DNA can even jump to another chromosome in the same organism where it reinserts itself. In some instances, there is replication of the transpondable element so that the replicated segment of DNA moves to another site where it inserts itself, while the original copy remains in place. Obviously, this type of mutation can add base pairs to the genome of the organism in which it occurs but these base pairs would be copies of pre-existing genetic material already existing in the genome. An interesting phenomenon occurs when a transpondable element inserts itself at a new site in the DNA molecule. Its presence at the new location influences the expression of genes in that location. This phenomenon can cause rapid changes, some of which are dramatic in the resulting DNA.[12] Transpondable element mutations will be discussed again in Chapter 12.

The same Barbara McClintock of whom we learned about in chapter six continued her research of maize and discovered the phenomenon of transposition, also known as transpondable element mutations or transposons, or jumping genes. She referred to the individual portions of chromosomes as "controlling elements" rather than the term gene. McClintock's work was about 30 years ahead of her counterparts in genetic research. Even though her work was not accepted at first, unlike Rosalind Franklin (Chapter 3), and Henrietta Leavitt (Chapter 8) who both died before any recognition could be accorded to their work, McClintock lived long enough to have her work assimilated into scientific practice. At 81 years of age, McClintock received the Nobel Prize in Physiology or Medicine for her discovery and description of the jumping gene.[13] Barbara McClintock was the fourth woman to receive the Nobel Prize in science.[14]

The phenomenon of transposition is thought by some scientist to play a significant roll in evolutionary theory because it can cause rapid changes to occur in a given species for natural selection to preserve or reject.[15] In addition, the notion that genes can move about in the genome of a given species, as described by McClintock, may have prompted Susumu Tonegawa, a Japanese molecular biologist working at Massachusetts Institute of Technology (MIT) to discover that genes of B lymphocytes (B-Cells) can rearrange the genes themselves. Tonegawa proved that 1,000 possible gene segments in B-Cells can rearrange themselves so that they can produce more than one billion specific proteins, one of which is an antibody needed by the immune system to fight off invading viruses or bacteria. In 1987 he received the Nobel Prize in Physiology or Medicine.[16] This phenomenon will be revisited again in chapter 12.

There are also other types of mutations known as chromosomal breakage. These mutations occur mainly during meiosis and rarely during mitosis. Sometimes segments of DNA become duplicated, translocated, deleted, or inverted. These mutations can happen spontaneously especially during meiosis, or from external causes, that is, from chemicals, viruses, or ionizing radiation. They are very serious. During meiosis, mistakes can occur during the process of crossing over.[17] These mistakes are analogous to a musician skipping over several bars or lines of the music and leaving them out altogether.

Or they could be like a musician skipping over several bars or lines of the score and reinserting them later into the rendition. Or it could be like skipping over several bars or lines and reinserting them back into the same place by playing the music backward. They also could be like a musician repeating several bars or lines of music by playing them over and over.

To help understand what the terms duplication, translocation, deletion, and inversion mean, study the next four diagrams. These mutations happen mainly in meiosis during crossing over and therefore are seen for the most part in sexually reproducing eukaryotes.[18] However, chemicals or viruses could also cause chromosomal damage. Stem and loop mutations and transpondable element mutations could also be involved in some cases of duplication, which could also occur in bacteria and single celled eukaryotes.

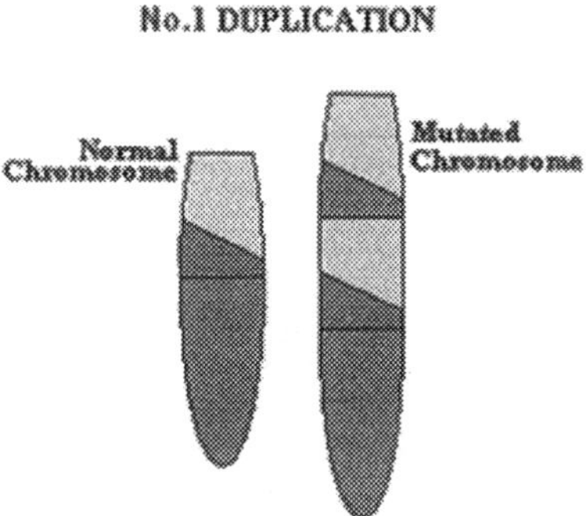

Figure 7-4

In the first diagram (Figure 7-4), you can see the duplicated upper portion of the normal chromosome in the mutated chromosome, like the musician playing several bars of music twice. A transposon could also be responsible for this type of mutation.

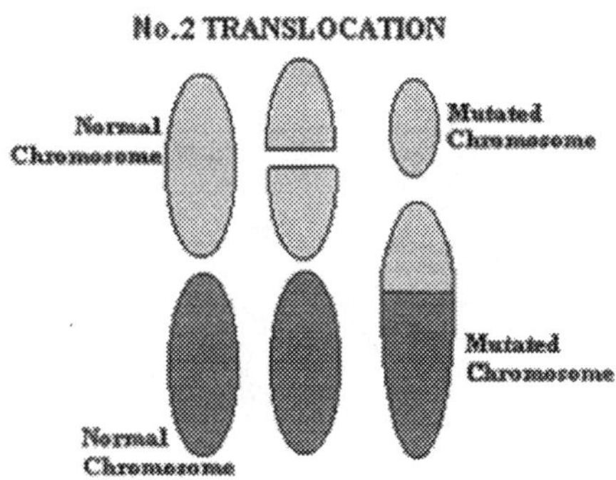

Figure 7-5

The second diagram (Figure 7-5) shows two chromosomes, one above the other. The upper chromosome breaks in two and part of it becomes attached to the lower chromosome. This is translocation. This would be like a musician taking part of one composition and inserting it into another composition. Most likely this kind of mutation would occur during crossing over in meiosis. Sometimes the divided chromosome remains divided, causing a triplet chromosome known as a trisomy. This usually happens during meiosis. In humans, this type of mutation results from translocation of chromosome 21 and is called Down's syndrome or trisomy 21 (formally mongolism). It causes a form of mental retardation[20] and is frequently associated with congenital heart defects. These babies tend to be born to older mothers.

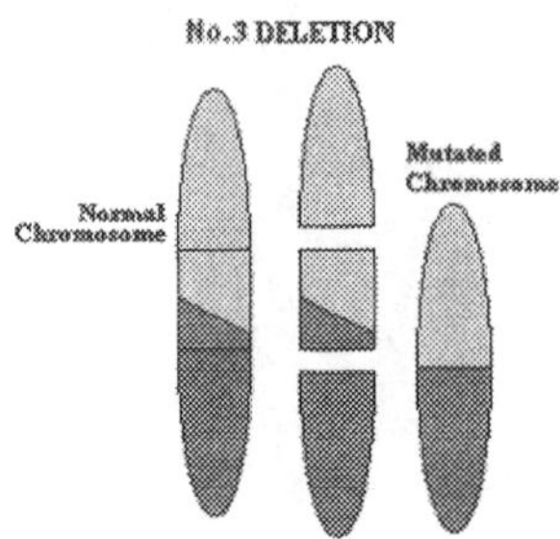

Figure 7-6

The third diagram (Figure 7-6) demonstrates a loss or deletion of part of a chromosome. You can see this demonstrated in the diagram. This one would be like having several bars of music left out. Besides accidents occurring during meiosis, stem and loop mutations could also subtract genetic information.

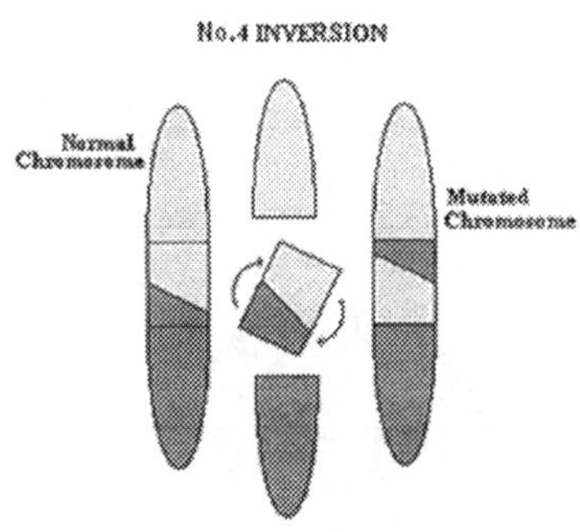

Figure 7-7

The fourth diagram (Figure 7-7), demonstrates inversion. This most likely happens during meiosis when a portion of a chromosome gets rotated and placed back in the same chromosome upside down. This would be like a musician playing several bars of music backward in the middle of a musical rendition.

A mutation introduced into the DNA in the first cell of a sexually reproducing eukaryote (zygote) will be present if not fatal in each succeeding daughter cell as the organism develops. The mutation can be introduced by either a mutated egg or sperm, or both. Either way, the resulting mutation will be present in the DNA of every nucleated cell in the new organism. If the original mutation is not fatal, each succeeding generation of this organism will most likely receive this mutation. External causes such as radiation, a chemical, or a virus can also attack and change a zygote's DNA, but because a zygote starts to differentiate soon after fertilization, there will be very little time for this to occur, and therefore this phenomenon will be very rare. A mutation occurring in a multi-celled sexually reproducing eukaryote after differentiation has proceeded to where the various tissues have developed, can only be passed on to new daughter cells of that tissue cell line. As previously stated, unless the mutation exists in one or both primordial germ cells, or the zygote, this kind of mutation cannot be passed to succeeding generations of that organism. In prokaryotes, a permanent mutation (that is, one affecting subsequent offspring of the mutated cell) can take place at any time from external causes but most likely would happen during DNA replication from internal causes. The same would apply with unicellular eukaryotes that don't sexually reproduce.

Another mutation, though very rare and most commonly seen in Caucasians, is cystic fibrosis. Children born with this mutation have a thick, sticky mucous secretion that they have difficulty clearing. These children are prone to frequent lung and airway infections, small bowel obstructions, pancreatic insufficiency, and cirrhosis of the liver. These result from a defective gene which codes for a protein 1,480 amino acids long. Just one codon for phenylalanine at position 508 is missing, which causes this protein to malfunction, resulting in the misery that these children must suffer.[21] This mutation occurs when an entire triplet codon is deleted.

Another mutation in humans that causes misery is hemophilia. This is a sex-linked disorder of the blood-clotting mechanism. Sex-linked means it is found predominately in one sex. Hemophilia occurs mostly in male children. There are several different kinds of mutations that make for a defective gene in Factor VIII in the blood-clotting cascade. Among these are point mutations, insertions, deletions, and amino acid substitutions, all of which cause a defect in the blood-clotting mechanism. Very small cuts will not stop bleeding. With this defect, the mutant person frequently may hemorrhage into joints after running or have internal bleeding. To maintain any semblance of normal living, these children must have frequent transfusions of human blood products that contain the normal Factor VIII.[22] The use of human blood products has exposed many hemophiliac children to AIDS and hepatitis C.

It stands to reason that at some time in our environment we become exposed to one or more of the external causes of mutations, such as chemicals, radiation, or viruses. Perhaps one of our cells sustains a mutation from one of the external causes mentioned above. What happens if one of these kinds of mutations occurs in a mature cell in your body? The answer to this question is the same as with other mutations. If this mutation is severe enough, it can cause that one cell to die. If that happens to one of your cells, the whole body doesn't die, only that one cell. Your body soon absorbs it, and you keep on living, never knowing that anything happened. However, instead of dying, some mutated cells start to replicate themselves autonomously and form tumors. Benign tumors expand locally and, to produce a cure, should be removed surgically. Malignant tumors, or cancers, can grow locally at their site of origin but also can spread to distant locations in the body by a process called metastasis. This is why it is so important to find cancers early in their development and have them removed before they spread.

In summary, mutations are the biological tailors that specialize in making accidental alterations in suits of DNA.

In the December 3, 1999, issue of *Science* (Vol. 286, No. 5446, pp. 1805-2032) is a series of four articles that deal specifically with the great lengths to which cells go to maintain the fidelity of the genetic messages found in the DNA. These four review articles are titled *Frontiers in Cell Biology; Quality Control*.

Here is a summarization of some of the salient features.

The high-fidelity mechanisms of cells prevent most mutations from occurring in DNA. This high fidelity also guards against mistakes that sometimes can occur during RNA or protein construction even when no mutation in the DNA exists.

Stella M. Hurtley's introduction to these articles states:

"Cells are the basic building blocks of living organisms, and the cell can be pictured as a very complicated factory of life. In order to maintain an effective internal regime and to prevent inappropriate attack by external factors, the cell needs quality control mechanisms to identify, correct, and prevent mistakes in its ongoing processes. The consequences of faulty quality control range from the cell death of neurodegeneration to the uncontrolled cell growth that is cancer."

Notice that last sentence of this quotation? *"The consequences of faulty quality control range from the cell death of neurodegeneration to the uncontrolled cell growth that is cancer."*

A prime example of neurodegeneration is the disease that Stephen Hawking has, called Lou Gehrig's disease. Its real name is amyotrophic lateral sclerosis. It got its nickname from the famous baseball player whose career and life were shortened by this tragic malady. The main problem is found in the spinal cord where the nerves that enervate the muscles start to die and scars form, which causes paralysis of the muscles and eventual death.

Cancer really begins when some cell in a given tissue in the body goes wild and starts multiplying rapidly so that it becomes a tumor. In both of these examples that the author gives, there is a problem of quality control in certain cells of the body. Mutations have occurred.

The first article is authored by Lars Ellgaard, Maurizio Molinari, and Ari Helenius. It describes many functions of the organelle known as the endoplasmic reticulum and how it works, *"A variety of quality control mechanisms operate in the endoplasmic reticulum and in downstream compartments of the secretory pathway to ensure the fidelity and regulation of protein expression during cell life and differentiation."* They also ensure that these proteins then get sent to their designated organelle. If a protein contains a mistake in it, the quality control mechanisms degrade it so that a protein improperly constructed never gets to participate in the cellular metabolism. One place where an improperly constructed protein would cause a problem would be a marker protein on the surface of the cell. Improperly constructed marker proteins would cause the immune system of the body to attack this cell as a foreign protein and destroy it or prevent other normal, positive cell to cell interactions. As you will recall from the discussion of protein synthesis, each protein is made, one amino acid at a time, in a long line. Then the long line of amino acids begins to fold on itself into a specific shape with specific characteristics that will allow it to function in a specific way. The endoplasmic reticulum provides a "special folding environment" that prevents release into the cytoplasm of improperly folded or improperly assembled proteins.

The second article of this series is titled *Posttranslational Quality Control: Folding, Refolding, and Degrading Proteins*. It was written by Sue Wickner, Michael R. Maurizi, and Susan Gottesman. It is noteworthy that it credits the proper folding of the protein to molecular chaperones and proteases.

When both of these quality controls fail, damaged proteins accumulate in the tissues and cause amyloid diseases. The abnormal protein, amyloid, is deposited into the tissues. The problem that the affected person has depends, to a large extent, into which tissues the amyloid is deposited, such as the liver, kidneys, or muscles. Especially problematic are the unwanted amyloid plaques deposited in the brain, implicated in dementia.

The third article, *Quality Control Mechanisms During Translation,* is by Michael Ibba and Dieter Soll. This article goes into considerable detail about how quality control is monitored during translation. While tRNAs and mRNA are being made, this process monitors for mistakes. Also, when the tRNA loaded with the appropriate amino acid approaches the mRNA attached to the ribosome, further monitoring takes place. Here is a quotation from the last sentence of the article's summary. *"Recent studies have begun to reveal the molecular mechanisms underpinning quality control and go some way to explaining the phenomenal accuracy of translation first observed over three decades ago."*

The last in this series of four articles has to do with the quality control by DNA repair. Tomas Lindahl and Richard D. Wood reviewed this subject. The preamble to this:

"Faithful maintenance of the genome is crucial to the individual and to species. DNA damage arises from both endogenous sources such as water and oxygen and exogenous sources such as sunlight and tobacco smoke. In human cells, base alterations are generally removed by excision repair pathways that counteract the mutagenic effects of DNA lesions. This serves to maintain the integrity of the genetic information, although not all of the pathways are absolutely error-free. In some cases, DNA damage is not repaired but is instead bypassed by specialized DNA polymerases."

These four articles in *Science* show why mistakes in protein construction and RNA translation are so rare and why mutations are rare events. A successful mutation, that is, one that is not corrected by the proofreading ability of DNA polymerase I and other DNA repair mechanisms mentioned above are, for the most part, very rare. Many are devastating, but some cause only minimal or moderate disturbances. So would it be possible that some mutations could be helpful or beneficial to the organism in which they occur? A beneficial mutation in a given organism has to be one that increases the chances for that organism to survive in its environment. From a theoretical point of view, this notion would have to be factual, not a value judgment.

Beneficial mutations are extremely rare events. An upcoming chapter will discuss possible beneficial mutations that have occurred in bacteria developing mechanisms to escape harmful effects of antibiotics, and some plants, mainly weeds that have developed methods to escape the harmful effects of herbicides and some insects that have developed ways to counteract the effects of poisonous sprays. When a beneficial mutation occurs in a bacteria, plant, or insect, these mutations are beneficial only to the organism containing them but not to humans who constantly battle bacteria, weeds, and insects. But please note that the bacteria stay bacteria; the plants stay plants, and insects stay insects. Another kind of beneficial mutation is called a suppressor mutation. This type of mutation corrects a previous mutation in the DNA bringing it back to normal function again.

The most important principles to remember from this brief study of mutations are that all mutations are rare, random, events as to where they occur in the DNA and when they occur. Silent mutations, neutral mutations, and deleterious mutations though rare are the most common. If beneficial mutations occur at all, they are extremely rare events.

SUMMARY

1. A mutation is any change that is accidentally made in the DNA from an original. Mutations that get passed the correcting effects of cellular quality control mechanisms are quite rare. Each is a random event, as to where in the DNA it takes place and as to timing when it occurs. Mutations may be as small as changing only one base pair or as large as splitting an entire chromosome.
2. One of the most frequent as well as simplest of mutations is the base pair or point mutation. It gets its name from the fact that only one base pair is changed with each of these mutations. It comes in four different categories based on the effect that the mutation produces. They are silent, neutral, nonsense, and missence:
    a. Silent base pair mutations produce no change in the amino acid specified in the DNA. This is because one base pair in a codon is exchanged for another which codes for the same amino acid.
    b. Neutral base pair mutations produce minimal change because even though the mutation changes one codon to code for a different amino acid, the different amino acid acts similar to the displaced one so that the resulting protein continues to function normally.
    c. Nonsense base pair mutations usually cause major disruption because the resulting codon is changed to a stop codon. The resulting protein may not function at all.
    d. Missense base pair mutation changes a given codon to code for a different amino acid which functions in such a way that the resulting protein functions poorly or not at all.
    e. Frame-shift mutations result when one base pair is added or one base pair is subtracted from the original DNA. Because the ribosome translates the mRNA one

triplet at a time, the added or subtracted base pair with this mutation causes the next triplet that the ribosome translates to be composed partly from the preceding codon and partly from the succeeding codon. Each succeeding triplet in the gene continues to be translated in this way until the end of the gene is reached.

3. There are many other kinds of mutations that usually mutate much larger sections of DNA.They include stem and loop, duplication, translocation, deletion, and inversion. Most of these occur during meiosis and therefore are found for the most part in sexually reproducing eukaryotes.
4. There are two basic causes of mutations:
   a. External causes are viruses, certain chemicals, and ionizing radiation such as ultra-violet light, X-rays or cosmic rays.
   b. Internal causes are mistakes made during DNA replication, either during mitosis or meiosis.
5. There are three general types of mutations categorized by their effects, or lack thereof, on the organism in which they occur:
   a. Neutral mutations or silent mutations have already been described above.
   b. Deleterious mutations are of two types: those that hurt the organism, but not enough to kill it (missense) and those that are so severe that they kill the organism outright (nonsense).
   c. A beneficial mutation theoretically must help the organism in which it occurs to improve its chances to survive. These mutations, if they occur at all are extremely rare.
6. Multiple repair mechanisms present in the cell also corrects many mutations on the spot, so to speak, making mutations even rarer.

Chapter 8

How Much Time

In Chapter 1 we concluded with Isaac Newton who believed that the universe was infinite in size and Kant who believed that it was infinitely old. Because infinity both as to size and age has no boundaries, this chapter will show that the universe had a beginning and possibly will have an end. It also has boundaries as to size as well. So how old and big is the universe, and how much time has elapsed? We'll start with Albert Einstein and what he discovered.

Albert Einstein[1] (1879-1955) was born in Ulm, Germany, of Jewish parents. They soon moved to Munich where he grew up. When he was only age five, his father gave him a compass with which he became thoroughly intrigued. His early education began in a Catholic elementary school. At age ten, he entered the Luitpold Gymnasium (a kind of German middle school and higher), to continue his education. At age twelve, an uncle introduced him to the wonders of Euclidean geometry by explaining to him the Pythagorean Thrum. Einstein had trouble adjusting to the Gymnasium, and at age fifteen was advised to leave before receiving a diploma. Some thought he would never amount to anything. However, Einstein taught himself, continuing to read books on physics. When he was 17, he was able to enter the Polytechnical Institute at Zurich, Switzerland. However, had it not been for a friend, Marcel Grossman, taking notes for himself in class, Einstein would have flunked, as the courses bored him. Einstein studied Grossman's notes so he could pass the examinations.

After graduation, Einstein could not secure an academic position due to the fact that he had antagonized his professors so much that they would not give him good references. Because of this, he scarcely was able to eke out an existence with odd teaching jobs here and there after leaving the Polytechnical Institute. It was during this time that he became a citizen of Switzerland. His old buddy, Marcel, came to his rescue again by helping him obtain a job as a junior patent examiner in the Swiss Patent Office in Bern, Switzerland. Now, with a secure income and time on his hands, he was able to pursue his interest in theoretical physics. At age 26, during the year 1905, he not only received a doctorate from the University of Zurich with a theoretical dissertation on the dimensions of molecules, but in addition he authored four separate, revolutionary scientific papers, all of which were published. This set of scientific quintuplets, fathered seemingly so effortlessly by a man in his mid-twenties, set the stage for much of physics through the twentieth century and beyond. Unlike Newton, Einstein was able to get his scientific discoveries published. But like Newton, this brilliant streak of genius came to fruition in his mid-twenties.

The first paper explained the photoelectric effect of light striking certain metal surfaces. Einstein conceived the idea of light traveling in individual wave packets rather than in continuous waves. These individual packets of light energy can produce an electric current when they strike metal and jolt electrons out of their orbits. We now call these packets photons. This was a revolutionary idea of categorizing light waves. For this, sixteen years later, he won the Nobel Prize in physics in 1921.

The second paper was an explanation for Brownian movement of small particles suspended in a gas or liquid. Einstein explained that the jerking motions result from being bombarded by atoms or molecules invisible with a microscope. When the smaller, invisible particles strike the visible particles with enough momentum, they jostle the larger particles visible under a microscope, causing them to move chaotically.

The third and fourth papers formed the basis for his special theory of relativity. He deduced the entire special theory based on his accepting the results of the famous Michelson-Morley experiment, performed in 1887, which other scientists had either questioned or rejected. This experiment was named after the two men who had performed it. (Michelson is famous for his measurement of the speed of light. In 1880, he measured the speed of light at 299,910 kilometers per second or approximately 186,000 miles per second.) Several years before Michelson and Morley performed their famous experiment in 1887, Clerk Maxwell proposed the existence of a substance he called *aether* that he thought filled the vast regions of space. He believed light waves needed a substance in which to propagate just as sound waves need air in which to travel. Many scientists in the latter half of the nineteenth century believed that *aether* was motionless in space and that the Earth traveled through this motionless substance just as an automobile travels through air on a windless day. However, just as all children know if they stick their hands out the window as the car travels through motionless air, an artificial wind is encountered as the car speeds through the air around it. The Michelson–Morley experiment was designed to measure the speed of the Earth as it traveled through the proposed substance *aether*.

This famous experiment measured the speed of light at various points as the Earth turned on its axis. Michelson and Morley designed it to be able to detect small differences in the speed of light as it traveled through the *aether* that was supposed to exist everywhere in space. If the proposed substance *aether* existed, the speed of light should vary as measurements were made at different points of reference. From these slight differences in the speed of light at different points of reference, the speed of the Earth could be calculated as it traveled through the hypothetical *aether*. The experiment failed to detect any differences in the speed of light at the various points of reference.

The Michelson-Morley experiment proved that there was no such substance as *aether*. For the scientists of the late nineteenth century, the results of the Michelson-Morley experiment posed what appeared to be an insurmountable paradox. Einstein solved the problem by rejecting the existence of the aether and his unique perspective of time. He came up with the notion that the passage of time is different for different observers who have different frames of reference. In other words, the rate at which time flows can be faster or slower for different observers. He also deduced that the speed of light was the same when measured by any observer, no matter what his speed was in relation to the light source.

Let us illustrate. Suppose you are standing beside a railroad track with a device to measure the speed of a train or an object on a train. One empty flat car passes by at ninety miles an hour. At the back of the car, a big league pitcher throws a baseball toward the front of the car at the rate of 90 miles an hour. Your measuring device would read the speed of the baseball as 180 miles an hour. This would represent the speed of the train plus the speed of the ball. However, when you reverse the experiment, the outcome is different. If the pitcher throws the ball at the rate of ninety miles per hour from the front of the car toward the back of the car as the train passed you at ninety miles per hour, the ball would appear suspended in air. Your measuring device would register zero miles per hour for the baseball. This thought experiment works fine with baseballs but does not work for light. If a beam of light is substituted for the baseball, a measuring device on the train and one on the ground would measure the same speed for the light ray. This experiment works regardless of how fast the train is going or from which end of the car you direct the beam of light.

For nearly two decades, eminent scientists tried in vain to solve this paradox. In 1905 Einstein, the 26-year-old high school drop-out, solved the problem with the novel idea that the rate of time flow is different for people traveling at different speeds. In other words, the rate of the passage of time is not absolute. The clocks of observers traveling at different speeds relative to each other keep different time. For example, take two synchronized clocks, keep one on Earth and place the second clock in a spaceship. If we send the spaceship on a long space journey, at speeds approaching the speed of light, when it returns, its clock will have run slower than the one on Earth. The faster the spaceship accelerates, the slower the clock will tick. Astronauts who ride in our present spaceships do not notice any differences in the timing of their wristwatches than those on Earth because they never approach the speed of light.

Starting with this premise, Einstein, through a series of mathematical manipulations, was able to devise his now-famous equation, $E=mc^2$. E stands for energy, m stands for the mass of any object, and c stands for the speed of light. From this one equation have come the atomic bomb, nuclear reactors, the hydrogen bomb, and the still-unsuccessful attempts at making a fusion reactor. Fallout from all of this is around us everywhere, no pun intended.

We know the ideas that produced the equation $E=mc^2$, as the theory of special relativity. Einstein was not content to stop with this accomplishment, but kept right on working with his equations. His idea was to unite these equations with the laws of gravity. Again, he used a thought experiment to help him find the answer. He considered having a room, like an elevator, out in space so far away from any massive body, such as a star or galaxy, that the gravitational force for practical purposes was zero. A man floating inside the elevator would not know up from down or sideways. Suppose a force applied to the "top" of the elevator began to move it in a certain direction. If it accelerated at a uniform rate, the man inside the elevator soon would find himself standing on the "floor" of the elevator. He would not be able to tell the difference between acceleration at a uniform rate or whether his room was hanging suspended from a tree limb in a gravitational field.

Einstein also used his idea of the elevator in space to show light is bent by gravity. If the elevator had a small window in one side through which light from a distant star could penetrate, the beam of light would travel across to the other side. However, as it was in the process of traveling to the other side of the elevator, the elevator would have moved "upward." Therefore, the projection of light against the side of the elevator opposite from the window would appear to be lower than the window through which it had entered. To the person inside the elevator, the light ray would appear to have been bent. But if he thought that it might be hanging from a tree in a gravitational field, he would think that gravity had bent the light.

Einstein derived the "Principle of Equivalence" from these thought experiments. This principle maintains that it is impossible to tell the difference between the pull of gravity and uniform acceleration. It took him a decade, from 1905 to 1915, to work out all the details, but he was finally able to bring the laws of gravity into the arena of relativity. He then brought forth what is known as the general theory of relativity.

Though the math is too complicated for us to consider in this book, the basic ideas are easy to understand:

1. Space is curved, which means that it has an edge. Although it is large, it is finite.
2. Space is expanding.
3. If space is expanding now, it implies that at sometime in the distant past it must have been all together in one small place. This implies that the universe had a beginning.

This implication of a beginning for the universe was distasteful to Einstein because the prevailing wisdom depicted the universe as infinite and static. But in 1915, Einstein could not believe what his math had found. A decade earlier, at age 26, based on the results of the Michelson-Morley experiment, he had found it easy to extricate himself from the paradigms of *aether* and *absolute time* handed to him by his scientific predecessors. He replaced it with the almost-unthinkable idea that time was not absolute and that *aether* did not exist. Now, at age 36, he could not believe that the space of the universe was a large but finite expanding volume. To cling to his mathematics and the ideas of infinite space and time (the paradigms of Newton and Kant), he paradoxically introduced a purely empirical number into his equations that he made up out of his head. He called it the cosmological constant. In his mind, the universe became infinite and static when he added this to his equations.

On the other hand, to give credence to his new general theory of relativity, Einstein thought of two ways to prove its correctness.

1. He said that the perturbations of the orbit of the planet Mercury would obey the principles of his general relativity better than Newton's laws of motion combined with his laws of gravity. His equations proved to be more accurate than Newton's with regard to predicting Mercury's orbit.

2. He said that observations of starlight would reveal that light rays bend as they pass by our sun, when in full eclipse by our moon.

An Englishman named Eddington traveled to an island off the coast of West Africa in 1919 to view a total solar eclipse. Observed in the totality of the eclipse, light from a nearby star did bend as it came around the sun. With the sun in full eclipse, Eddington found that the light from a distant star was bent less than two arc-seconds as it came around the sun. An angle of two arc-seconds is not much of an angle to measure; in fact the angle is only 1/1800 of a degree.[2]

The measurements obtained by Eddington proved to be quite fortunate for Einstein and his general theory of relativity. Eddington's measurements might just as easily have proven Einstein wrong as right. If there had been any flaws in his equipment, this might have dealt a serious blow to Einstein's general theory. On the other hand, Eddington might have, subconsciously, skewed the very small angle he measured in Einstein's favor. In the recent past observations involving the gravitational bending of light have been observed confirming Eddington's original observation. In *The Brief History of Time,* Dr. Stephen Hawking says that in the history of science the interpretations of experiments have often been skewed toward the desired outcome.[3] As an example, two scientists in Utah thought they had discovered a method of cold fusion. As it turned out, other scientists were unable to reproduce their measurements. The scientists in Utah had misinterpreted the outcome based upon what they hoped the outcome would be, and they fell into severe disrepute.[4] Their beautiful idea was killed in the scientific womb like a therapeutic abortion.

Alexander Friedmann was a Russian meteorologist also interested in mathematics. In the early 1920s, he obtained a copy of Einstein's *general theory*, published in a German science journal. He digested it carefully, including Einstein's cosmological constant. To his surprise, Friedmann found a simple algebraic mistake in Einstein's calculations. When corrected and when the cosmological constant was removed, it showed that the universe was expanding. It is now thought Friedmann probably was the only scientist in the early 1920s, including Einstein himself, to believe what Einstein's general theory had predicted: that the universe was expanding. In Freidmann's mind, there were two apparent possibilities:

1. If the average mass density of the universe was less than a certain critical amount, the expansion would continue forever.
2. If the average mass density of the universe was more than a certain critical amount, the expansion would gradually slow down until it came to a complete stop. At this point, it would start contracting until it came crashing in on itself.

Friedmann predicted what Edwin Hubble was going to discover not long afterward: the universe is expanding.[5] The algebraic mistake and the cosmologic constant placed obstructions in Einstein's mind for the "normal delivery" of his theory to the world. Nevertheless, the pregnancy test was positive as noted by the mathematics concerning the orbit of Mercury, and bending of light around an eclipsed sun. Friedman made the diagnosis and surgically removed the algebraic mistake in utero before birth. Hubble later delivered the theory by Caesarian section with his discoveries and confirmation of an expanding universe. We must credit Einstein with delivery of the metaphorical placenta--the cosmologic constant--which he removed a decade and a half after he added it.

The third of seven children, **Edwin Hubble** was born in Marshfield, Missouri, November 20, 1889. Soon after his birth, his family moved to Evanston, Illinois, and about two years later moved again to Wheaton, Illinois. At Wheaton High School, Edwin distinguished himself as a scholar as well as an athlete. He entered the University of Chicago at age sixteen, earning high grades in mathematics, chemistry, physics, astronomy, and languages. Even though he was about two years younger than most of his classmates, he won letters in track and basketball.

After graduation from the University of Chicago, he received a Rhodes scholarship and continued his studies in England, at Queens College of the University of Oxford. He studied jurisprudence, apparently because his father and grandfather wanted him to become a lawyer. After completing his studies in England, he returned to the United States and soon "chucked" the legal profession to pursue a life of astronomy. As a result, he contacted one of his former astronomy professors at the University of Chicago, who helped him become a graduate student in astronomy at Yerkes Observatory in Williamsbay, Wisconsin, obtaining a scholarship that covered tuition and living expenses.

In August 1914, while a student, Hubble attended the American Astronomical Society's meeting at nearby Northwestern University, when Vesto M. Slipher, from the Lowell Observatory in Arizona, initiated a cosmological controversy (more birth pains) with his observations of nebulae. Nebulae was a catch-all term used to designate cloudy objects noted in the heavens. He was the first person to study these distant celestial objects. Now we know these nebulae to be galaxies or island universes like our Milky Way. Slipher had attached a small spectroscope to the Lowell telescope and noticed that the spectral lines shifted toward the red end of the spectrograph when he directed the telescope toward the various nebulae.

From this, he deduced that these objects were traveling away from Earth at very rapid speeds. The more distant the nebulae appeared to be, the more rapidly they appeared to be receding from Earth. Although Immanuel Kant had proposed in 1755 that some of these cloudy objects might represent "island universes" similar to the Milky Way, the idea was difficult to prove or refute. Apparently, this notion sparked Hubble's mind with the idea to pursue research along this line. The title of his doctoral dissertation was "Photographic Investigations of Faint Nebulae."

In October of 1916, Hubble contacted George Hale, the director of Mt. Wilson Observatory in Pasadena, California. At that time, Hale was trying to put together a staff for the 100-inch reflecting telescope the observatory was building. He offered Hubble a job, subject to the completion of his Ph.D. degree.

About one month after receiving his docterate in May, Hubble reported for duty to the US Army. The United States entered World War I on April 6, 1917. Hubble became a battalion commander of the 86th Black Hawk Division. In September 1918, Hubble's division was in Europe, but the signing of the armistice in November kept him from combat. After his discharge from the army, Hubble took up his duties at Mt. Wilson in August 1919, becoming one of the first researchers to use the 100-inch telescope.[6]

Over the next decade, Hubble used ingenious methods of research to show that many of these so-called nebulae were island universes. His tests proved they were all receding away from Earth at ever-increasing rates of speed. The farther the nebulae (or galaxies) were from Earth, the faster their speed seemed to be. This observation, accomplished by studying the red shift of light waves as they traveled across the vast distances of space, revealed that the universe was actually expanding in all directions. But expanding from where? The implication was that the universe had expanded from a central point and had its beginning there.

Hubble's discovery proved Einstein's general theory was right, just as Alexander Friedmann had predicted. Hubble published these data in 1929. In 1931, Einstein conceded that the universe was not infinite and static, but was, in fact, finite and expanding. He withdrew his cosmological constant at that time, stating that it was the biggest mistake of his life.[7]

To better understand how Slipher and Hubble proved that the universe is expanding, a short discussion about spectroscopes and spectrographs is in order. So, what are spectroscopes and spectrographs?

Sunlight and starlight, though appearing to be white light, are actually a mixture of all of the colors of light. If sunlight passes through a triangular-shaped piece of glass, called a prism, the individual colors become separated. An artificial miniature rainbow forms with the red color on one side and the blue color at the other. These separated colors form a spectrum. A spectroscope is a machine that allows light from a given source to pass through a slit that is then focused onto the prism by a couple of lenses (the more modern spectroscopes use a grating rather than a prism). After the light has passed through the prism, which separates the colors, another set of lenses is used to focus the light again for visualization. The name of this machine is a spectroscope. When a camera is substituted for the observer's eye and photographs are taken, the photographs are called spectrographs. A very sophisticated spectroscope can measure the wave length of the light passing through it. The wave length of the light at the red end of the spectrum is longer than the wave length at the blue end. When a light source is speeding away from the observer, the light waves become stretched so they shift toward the red end of the spectrograph. The faster the objects are traveling away, the more the light will shift. This is known as the red shift. However, when a light source is moving toward an observer, the light waves will be compressed which will shift them toward the violet end of the spectrum, the opposite of the red shift. It is only necessary to remember that a spectroscope, when attached to a telescope, is a sort of a celestial speedometer.

HENRIETTA LEAVITT

Edwin Hubble's discoveries might not have been accomplished had they not been preceded by a discovery by Henrietta Leavitt. She was a graduate of Radcliff College who came to work as a volunteer at Harvard College Observatory. She started this work in 1896 but was interrupted by several years of illness, which caused her to become partially deaf. However, in 1902, she returned full time to Harvard. She was assigned the tedious and boring task of cataloging variable stars. Variable stars vary in brightness or luminosity from time to time. Among the many kinds of variable stars are those called Cepheid variables. The time that it takes for a Cepheid star to vary from its brightest light to its faintest light and back again is called the period of a Cepheid star. Each Cepheid has its own repeating time cycle.

Leavitt had to compare the variation in brightness of a star on two photographic plates taken at different times. All of these photographs were of stars in the Magellanic Clouds. By studying these photographs carefully, she found 1,777 new variable stars, which also included more than 20 Cepheid stars. She discovered the brightest Cepheid stars have longer periods of variability than Cepheids with less brightness. The Magellanic Clouds can be viewed only from locations in the Southern Hemisphere and are small companion galaxies to the Milky Way. They were named after Magellan who was the first European to sail around the southern tip of South America. The photographic plates that Leavitt studied were photographed in Peru.

When the distance from Earth to a relatively close Cepheid star that can be measured by triangulation is compared with another one of much greater distance, the distance to the second Cepheid star can be calculated using Leavitt's discovery. The period-luminosity relationship she developed became the most accurate method available to astronomers of calculating distances in the universe. This little-known woman from Harvard produced the method by which Hubble was able to calculate the distance to other Cepheid stars in other galaxies. At a relatively young age, she died of cancer before any significant recognition could be accorded her. The delivery of her scientific offspring in 1912 was adopted by Ejnar Hertzsprung a year later, when he calculated the distance of some relatively nearby Cepheid stars.

Using Leavitt's discovery as a basis for his research, Hubble was able to calculate the distance to the Andromeda galaxy, our nearest galactic neighbor.[8] It helped greatly that for the first time individual Cepheid stars in our nearest galactic neighbor could be seen with the new 100-inch telescope that Hubble was using. Before this, distant galaxies appeared like clouds of light blurred together and individual stars could not be identified. With the larger telescope, Cepheid stars in Andromeda could be identified and, because of Leavitt's work, Hubble could calculate the distance from Earth to Andromeda.

Through most of the 1920s, Hubble did his research and everywhere he looked, all galaxies were found to be moving away from Earth. Hubble confirmed what Slipher had first noted with his little 30-inch refractor connected to a spectroscope. In 1931, 16 years after Einstein first put forth his cosmologic constant, he was forced to remove it because of Hubble's research. It is interesting to note that Henrietta Leavitt's little-known discovery indirectly brought correct understanding of Einstein's theory of relativity, even to Einstein himself.

ENTROPY

Heat is measured in degrees of temperature on a thermometer. Though various thermometers were invented in the past, at this time there are basically two, Fahrenheit and Celsius. The distance between the degrees on the Fahrenheit scale are closer together than those on the

Celsius scale. On the Fahrenheit scale water freezes at 32° and boils at 212°. However, on the Celsius scale water freezes at zero degrees and boils at 100°. For the most part, scientists use Celsius thermometers. In addition to Celsius there is another method of measuring temperature called Kelvin, or K. The markings on a Kelvin thermometer would be exactly the same distance apart as those on the Celsius thermometer; however, zero on the Kelvin thermometer would be minus 273.15 degrees below the zero on the Celsius thermometer. Minus 273.15 degrees on the Celsius scale is considered to be absolute zero. There can be nothing colder than that. In fact there is no thermometer that can be made to measure down to absolute zero. The absolute zero on the Kelvin scale has been determined by other means.

We say that something is cold if it feels cooler than our body temperature. We say something is hot if it exceeds our body temperature. However, when something is cold as we perceive it, this actually means that it contains less heat than our bodies and when we say something is hot it contains more heat than our bodies. At absolute zero on the Kelvin scale there would be no heat whatsoever left in the substance. The temperature of absolute zero has never been reached, nor can it be.

Now that we know how to measure temperature, let's now turn our attention to what heat is and how it can be transferred from one place to another. Heat is simply the degree to which atoms or molecules of a substance bounce around hitting each other. The colder an object is, the less the atoms or molecules are bouncing back and forth against each other. The hotter an object, the greater is the intensity with which the atoms or molecules of that object are reverberating back and forth colliding with each other. At absolute zero all the atoms or molecules of a substance would be absolutely still.

Heat can be transferred about by three means: conduction, convection, and radiation. As an example of conduction, take a long metal rod and place one end in a flame. After a time the whole rod will become heated, even to the end that is not in the flame. This is because as the atoms or molecules in the heated end start to increase in the intensity of their bouncing about, they convey their energy to adjacent atoms or molecules and so on to the other end of the rod. This is conduction.

Convection is demonstrated by a wood stove in a cold room. When a fire is started in the stove the air around the stove becomes heated, causing the heated air to rise. As a result more cool air near the floor and by the stove moves to take the place of the heated air that already rose towards the ceiling. The colder air becomes heated and it too rises. As the heated air moves toward the ceiling it starts to cool again and begins to settle back towards the floor only to be heated again by the hot stove. Soon convection currents are set up in the room with heated air continuously moving upwards and away from the stove only to settle down again as it cools followed by another warming. This is an example of convection.

The heat from the sun is an example of radiant heat. Infrared and other electromagnetic waves travel through empty space from the sun to our planet earth. When they strike an object they are converted to heat as when you feel the warmth generated by the sun's rays striking you when standing in the sunlight on a cold morning. By radiation of electromagnetic waves, energy can be transferred from one place to another. This is not confined to the sun's rays. The same phenomenon can be perceived when you stand near a hot stove. The infrared rays emanate from the stove. When they strike your body you feel the heat that they generate.

Now that you know what heat is, how it moves around, and how to measure its intensity with a thermometer, you are now ready to understand the first two laws of thermodynamics. The two laws of thermodynamics provide another way to prove that the universe had a beginning. The first law states that, although energy can be changed from one form into another, it can never be created or destroyed. Therefore, the total amount of energy in all its different forms, present in the entire universe, has remained constant since it came into existence at the instant of the Big Bang. The second law states that energy can be transformed only from a higher state into a lower state, never the reverse, and with each transfer some energy becomes unavailable for work. That portion of energy that has become unavailable to perform useful tasks has dropped to a state that is too low to be used for work. It can never be recovered again. This wasted energy is called entropy. Entropy is a measure of energy unavailable to do useful work.

To quote Gary F. Moring on page 69 of the *Complete Idiot's Guide to Understanding Einstein*, he says, "It's not that we're using up energy, it's that we're transforming it into useless forms. Only some of it is used; the rest of it is wasted." Energy comes in many forms, mechanical as with a wound-up clock; chemical as gasoline needed to run a car; atomic to power a submarine; electric to light a lamp; and radiant energy that shines from the sun to warm the Earth. These are some of the most common examples of energy that we encounter almost daily. In each case, when any kind of energy is changed from one form into another, the transfer is never 100% efficient. Some of the original energy, always a smaller portion, is used to perform some type of useful work. The remaining energy and always larger portion is lost in the transfer and ends up in the reservoir of entropy; useless; incapable of performing any work. However, the total amount of energy has remained constant because the sum of the energy available that can perform work plus the energy lost to entropy remains the same.

Heat is one type of energy. It can move only from hotter places to cooler places, never the reverse. As an example, energy transfer occurs when heat from a stove moves to a cooler skillet to warm it, and from there into some eggs and potatoes to cook them. However, some of the heat warms the stove, some warms the side and handle of the skillet, and some leaks around the skillet and warms the room, even on a hot day. All the heat that did not cook the eggs and potatoes is wasted. It can never be recaptured again to do useful tasks. The heat that warmed the room cannot be recaptured to cook more eggs. It has entered the reservoir of entropy from which it can never escape. On a cold day the heat that leaked around the skillet and warmed the room could be considered useful to someone working in that room. However, eventually this heat becomes so diluted that it has no useful function and is absorbed by the reservoir of entropy.

A school bus provides us with another example of how entropy grows bigger with every energy transfer. Of the total amount of chemical energy in the diesel fuel used by the bus, only a small fraction actually moves the bus down the road. The excess heat produced inside the engine is wasted on purpose by the radiator to prevent the engine from burning up. However, on cold days some of the heat generated by the engine can be used to heat the inside of the bus. In addition to the radiator, a lot of the heat is also wasted via the exhaust pipe, to say nothing of the heat lost to friction of turning gears or the tires against the road. A larger amount of energy is always expended than is used to accomplish the desired work. The wasted portion ends up in the lake of entropy. Energy transfers needed to perform useful work are never 100 percent efficient. In fact, even with the most efficient systems devised, only 30 percent to 40 percent of the total energy expended is used for work and 60 percent to 70 percent goes to entropy where it becomes unavailable forever to perform useful work. The exhaust from an engine cannot be recollected to run the engine again. This unavailable energy is never destroyed; it does not disappear, but only moves to a lower state where it becomes unavailable to perform work. Eventually the unusable energy simply goes to warm the universe. To emphasize again, this energy existing in a lower state which cannot be used to perform work is called entropy.

PHYSICAL SYSTEMS

Closed Physical Systems

A closed physical system is a system where energy in any form cannot enter, nor exit from the system. Actually, a closed system is imaginary because no such system actually exists unless it is the universe itself. (This will be discussed later.) The best example of a nearly perfect closed physical system is a thermos bottle. It is a bottle within a bottle, usually made of glass, where the space between the two is a vacuum. Such a vacuum prevents almost 100 percent of heat transfer via convection and conduction. Radiation across the vacuum is diminished by having the surfaces on the vacuum side of each bottle silvered. This reflects almost all radiation from the outside of the thermos bottle back to the outside and almost all radiation arising from the inside of the bottle, back into the bottle. A minor leak in the system exists where the cork is applied at the top. If hot coffee is stored inside, this very small leak will allow some heat from the hot coffee to escape to the outside, which will allow the coffee to begin cooling. Or the reverse, if ice cold lemonade is inside, heat from the outside on a hot day will start to leak into the bottle through the outlet at the top, allowing the lemonade to start warming up. Regardless, the entrance or exit of heat is so minimal that the contents of the bottle stay very near the desired temperature for several hours. However, it is not a perfectly closed physical system. Theoretically, if a perfectly closed physical system existed, entropy could only increase, never decrease inside it.

Open Physical Systems

In an open physical system, energy of any kind can enter or exit the system as long as it follows the second law of thermodynamics which only allows energy to transfer from a higher state to a lower state and with each transfer some energy is wasted as it enters the reservoir of entropy. A campfire burning on a cold night in a desert is a good example of an open physical system.

Most of the heat goes up in the smoke and ends in entropy. However, some heat released from the burning wood warms the rocks surrounding the fire, and some can roast marshmallows or bake potatoes. When heat moves from the hot fire to the colder rocks, then after the fire goes out, heat moves outward from the rocks to warm the surrounding air until they cool down to the atmospheric temperature. The next day when the sun comes up, the surroundings begin to warm up. By noon the rocks around the fire pit have become warm again even with no fire. In each situation heat moved from a higher state to a lower state. Almost all the energy in each case was lost to entropy. Heat entered and exited in this situation following rules of the second law.

Confining Physical Systems

While it is true that there are only two kinds of physical systems – open and closed – there is a type of system in-between, which I shall call a confining physical system. Recall that a closed physical system does not allow any energy to enter or leave the confines of the system and that an open physical system has only two requirements with regards to the flow of energy either in or out of the system, namely those imposed by the second law of thermodynamics. The confining physical system is a variant of an open system in the sense that energy can enter the system or escape from it under specific confining and directing conditions. The most important characteristic of a confining physical system is its ability to control the intensity and give direction to forces resulting from energy released inside it. For the most part, confining systems are engines of one kind or another, which produce some kind of work.

Confining physical systems are bound by the limits with which a given force can be controlled. An internal combustion engine is an example. Relatively small perfectly timed explosions of fossil fuel inside the combustion chambers make the engine run smoothly. However, if a large stick of dynamite was to explode inside one of the combustion chambers, the whole engine would be destroyed. This is because the force of the dynamite exceeded the strength of the engines confining wall, whereas the relatively small perfectly timed fossil fuel explosions do not exceed the limits of confinement. The force from each successive explosion is directed against a piston which forces it down, causing the connecting rod to turn the crankshaft. The turning crankshaft can then be used to produce physical work, which is moving a mass through a distance. Confining physical systems produce work more efficiently than a purely open physical system such as a campfire because a confining system decreases randomness by giving direction to a force released inside its walls. Confining physical systems direct forces in such a way as to produce some work and some entropy. More order is expended to produce a smaller amount of a different kind of order in all confining systems. The efficiency with which a new form of order is created from the decreasing order of fossil fuel will always be much less than 100 percent. In fact, it is usually less than 40 percent, the rest of the energy (approximately 60 percent or more) is wasted or lost to the reservoir of entropy.

Though not obvious at first, the second law of thermodynamics is also a predictor of disorder. When the chemical energy of the fossil fuel explodes in the confines of an internal combustion engine, the concentrated and organized molecules of the hydrocarbon fuel become disordered as the carbon and hydrogen atoms are oxidized in the fiery explosion and heat is released. Work is done by the energy expended, but most of the energy ends up in entropy with a smaller amount used for work.

The above description applies equally to a bacterium such as an E. coli or a huge engine as in a railroad locomotive. An E. coli's confining physical system takes in chemicals containing energy in a higher state, which the bacterium's many metabolic processes extract. The energy is released step by step by enzymes so that the oxidation reactions do not destroy the confines of its physical system. These enzyme pathways force the energy in the proper direction and intensity to maintain the bacterium's life. It excretes waste products with energy in a lower state following the two rules of the second law of thermodynamics, namely that energy can only move from a higher state to a lower state and that some of the energy is wasted, ending up in entropy. A railroad locomotive energized by fossil fuel oxidized (burned) in its engine uses its confining system to direct the force of the released energy to pull the train down the tracks. It excretes the exhaust in a lower state than was located in the fossil fuel that it burned. All of the unused heat in the exhaust ends up in entropy. It too follows the rules of the second law. In both cases a larger amount of order is changed to a smaller amount of a different kind of order or work, plus most of the energy ending up in disorder in entropy. Every living biological organism follows the same process and therefore each can be categorized as a confining physical system. This is true of humans, hamsters, hemlocks, or hippos, and fungi, foxes, or ferns, and gnats, gorillas, or guinea pigs.

Disorder results when the organized molecules in the food you eat are oxidized by your metabolism to produce energy for your muscles and all internal organs including your brain. The energy to run your eyes as you read this page also comes from this source. Heat from the oxidation processes is generated to keep you warm. But disorder results when carbon dioxide is expelled in your breath, or when fecal material is excreted from your gastrointestinal tract or urine from the urinary tract. More order has moved to disorder obeying the second law, and much of the energy is wasted going to the reservoir of entropy.

As pointed out by Stephen Hawking in his book *A Brief History of Time*, page 146, disorder has many more combinations than order has. He uses the analogy of many pieces of a jigsaw puzzle confined in a box. Obviously there is only one combination that will display the picture of the puzzle but there are many orders of magnitude more wrong combinations of the pieces when they are shaken up inside the box. If after every shake the box is opened, there will be displayed a different combination of disorder as the pieces randomly move around in the box. Although there is no scientific law that says that the pieces cannot randomly all come together in the right place with a given shake of the box, the chance for this to happen by purely random means is as close to zero as you can get. The obvious way to get the pieces together in the right configuration is by the expenditure of energy to run the eyes, brain, and muscles of the arm and hand needed to reach the one and only goal of perfect configuration. Even here the second law of thermodynamics is obeyed. The order gained by placing the pieces of the puzzle in the correct configuration required the expenditure of more energy by the person putting the puzzle together than the order gained by the puzzle. When any order is gained, more disorder is required someplace else. In other words, there is no such thing as a free lunch.

Suppose that the thermos bottle was actually a perfect closed physical system where no energy could enter or exit. As time went by the coffee molecules would break down moving from order to disorder in obedience to the second law. In a perfect closed system entropy can only increase, never decrease. The fact that the second law will not allow a reversal of disorder to order is also one of the reasons that time always moves from past to present to future, never the reverse.

Even though our sun burns fusion fuel, it cannot burn for an infinitely long time. The sun is radiating heat away from itself in all directions, most of which is lost in space or absorbed by planets and their moons. In time, all of its fuel will burn up and it will go out. Don't worry. That will not happen for several billion years. The amount of energy absorbed by Earth as light and other types of solar radiation such as infrared is extremely small when compared with the sun's total output. Our sun's energy also is not renewable. The unabsorbed energy emanating from the sun is huge. All unabsorbed energy from the sun enters the lake of entropy. Then think of this: Every star in every galaxy in the entire universe is doing the same thing. The unabsorbed, or unused, energy radiating from these stars, including our sun, is entering the lake of entropy from which it can never be retrieved. The universe, as a whole, has no renewable energy resources. So from where did the original energy of the universe arise?

The reservoir, or lake of entropy, is growing larger every day. This means that the total amount of energy available to perform useful work in the entire universe is decreasing, as more and more energy moves to this ever-growing reservoir. However, the universe does not heat up because it keeps on expanding. Remember that the total amount of energy in the universe is constant. In other words, available unused energy, part of which can be used for work and part of which will be wasted as entropy, plus the unusable energy already located in the lake of entropy, is and always will be constant in obedience to the first law and second law of thermodynamics. This means that the total amount of energy available for doing useful work is becoming less and less every second of every day. This also means that there was more energy available for doing useful work in the past than there is today. If the universe were infinitely old, all energy available for performing work would already have been used up and heat death already would have occurred. Using this logic, scientists are able to prove that the universe had a definite beginning. If the universe is slowly running out of energy that can be used for work, or "winding down," then it must have been "wound up," so to speak, in the past.

With the removal of the cosmological constant and the algebraic mistake, Albert Einstein's general theory also showed the same logic. Run backward toward zero time, his equations show that the universe had a beginning. Slipher and Hubble's research showed that the universe had a beginning, but through different means. Slipher and Hubble discovered that the universe is expanding in all directions. From this it can be concluded that if the universe is expanding, then there was a time in the distant past when everything in the universe was very close together.

We have discussed three ideas: Einstein's theory of relativity, Slipher and Hubble's cosmological discoveries of an expanding universe, and entropy. Each, individually and together, illustrates that our universe had a definite beginning. The discoveries of an expanding universe and the theory of relativity show that our universe, though very large, is actually finite in size and age. It is not infinite in size and age, nor is it static, as previously supposed by Newton and Kant. Entropy, on the other hand, only shows that the universe had a beginning but contributes little with regard to its size.

When taken together, these are some of the greatest achievements of modern science. Because of these discoveries and others that have followed, we now know that Earth is a small planet that travels around a relatively small sun. The location of that sun is on the inner aspect of an arm of a spiral galaxy and at just the right distance from the galactic center. Our galaxy is only one of billions of galaxies in the universe. The Earth is not the center of the universe and it had a beginning.

These are classic examples of how scientific theories or paradigms have been forced to change as new information became available. Copernicus, Galileo, and Kepler forced changes in Aristotle's paradigms, which had been held so rigidly by the educated elite of that time. Einstein's equations forced science to realize that time is not absolute, but relative. However, he would not give up Newton and Kant's paradigm of an infinitely large and infinitely old universe, (even though his general theory predicted such) until forced to do so by the discoveries of Slipher and Hubble. *We must always be willing to shift our paradigms when new information becomes available and never be so rigid as to hold a paradigm in priority over reality.*

The Differences in Rates of Time Flow.

Scientists, for years, have speculated about the age of the universe until information derived from the Wilkinson Microwave Anisotropy Probe (WMAP) Satellite made it possible to calculate its age. For nearly a decade after the WMAP's launch in June 2001, until communications ceased on August 20, 2010, this satellite made observations and measurements of the cosmos, which it sent back to Earth for scientists to analyze. This data provided the information needed to figure the age of the universe quite accurately. Its age is calculated to be 13.772 + or – 0.059 billion years.[9]

On May 14, 2009, the European Space Agency launched a satellite, known as the Planck Space Craft, with similar, but more sophisticated measuring devises than those used by the WMAP Satellite. According to its findings the age of the universe has been calculated to be 13.798 + or – 0.037 billion years.[9]

So what were some of the discoveries that preceded space exploration, which caused scientists to change their concepts of the universe? Until well into the 20th century, scientists believed that the universe had always existed in a static form, except for the planetary motions traveling around the sun and moons circling the planets, etc. However, it was Einstein's equation of general relativity, which he first elucidated in 1915 that should have caused him to predict an expanding universe. Instead, he refused to give up the idea of a static universe and added the cosmological constant to his equation to preserve his desired consept.[10] In 1915, Einstein apparently did not know that Vesto M. Slipher, had already discovered that nebula, later known as galaxies, the major visible components of the universe, were moving away from our location in space, or vies versa or both. This was first observed by him at the Lowell Observatory in Arizona. A few years later during the 1920s, Edwin P. Hubble with the new 100 inch telescope on Mount Wilson in California[11] confirmed Slipher's findings in more detail. The galaxies were moving away from each other, showing that the universe was expanding.

Both Slipher and Hubble used the red shift of light waves to prove that the galaxies are traveling away from each other. As our galaxy and another move apart, light waves reaching us from the neighboring galaxy are shifted towards the red end of the color spectrum. This is very similar to what happens to the pitch of a sound coming from the horn of a car that just passed your car going in the opposite direction on a two lane road. If the person in the other car honks the horn just as the two vehicles pass, the sound waves reaching your ears will be stretched, as the two cars travel away from each other, causing the pitch to be lowered compared to what it was when each car was adjacent. This phenomenon is known as the Doppler Effect. The same principle applies to waves of light and is called the red shift.[12] As our galaxy travels away from the light coming from another source in the cosmos, the light waves are stretched toward the red end of the light spectrum, proving that the two light sources are moving away from each other.

The data gathered by Hubble, regarding the red shift stretching of light, was so convincing that it forced Einstein, in 1931, to remove his cosmological constant from his math, claiming it to be "the biggest mistake of his life."[13] He could no longer conceive of the universe as being static, but expanding in all directions. This obviously meant that, in the distant past, all of the galaxies of the universe had been very close together, suggesting a beginning. The evidence for a beginning was very repulsive to Einstein. Regardless, in 1927, even before Einstein removed his cosmological constant, Georges Lemaitre, found evidence for proof of an expanding universe in mathematics he had derived from Einstein's equation of general relativity.

For more than a decade after the 1920s, progress remained slow in the scientific community with regard to searching for a way to explain how and when the universe began. Then in 1946, *Physical Review*, published a paper by George Gamow, postulating a hot beginning for the universe, which with cooling, chemical elements would be produced.[14] A few years later, *Nature* published a paper by two of Gamow's students, Ralph Alpher and Robert Herman, in which they made predictions of a possible background radiation left over from a hot beginning.[15] These concepts remained stagnant, however, until the 1960s, when Robert Dicke and P. James E. Peebles at Princeton University began toying with the notion of the universe beginning with a huge blast of hot radiant energy, with the possibly of some leftover radiation reaching Earth now.[16]

Dicke and Peebles also knew from $E = mc^2$ that if the universe began when hot radiant electromagnetic energy was changed into mass, the immediate result of this process would cool the temperature of the primordial universe precipitously. Even so, after the production of stable mass, cooling of any leftover radiation would continue from dilution of heat due to the expanding universe. The expansion would stretch the electromagnetic wavelengths, producing a decrease in their frequency and a drop in the temperature of the universe in proportion to its increasing size. Recall, this stretching was noticed first by Slipher and later by Hubble in the wavelengths of the red portion of visible starlight, hence the name the red shift. However, this stretching occurs though out the spectrum of electromagnetic waves, not just in red light. Therefore, Dicke and Peebles knew, if any of this ancient radiation arrived on Earth now, it would have been stretched to weak invisible microwaves. But if detected, it would help to prove a hot beginning for the universe.

About the time these two giant minds at Princeton were considering these cosmic possibilities, Arno Penzias and Robert Wilson tested a very sensitive microwave detector. They noted the presence of microwaves coming in from outer space, no matter which way the detector was directed. By accident, they had discovered the radiation left over from the birth of the universe.[17] It became known as the Cosmic Microwave Background or CMB.

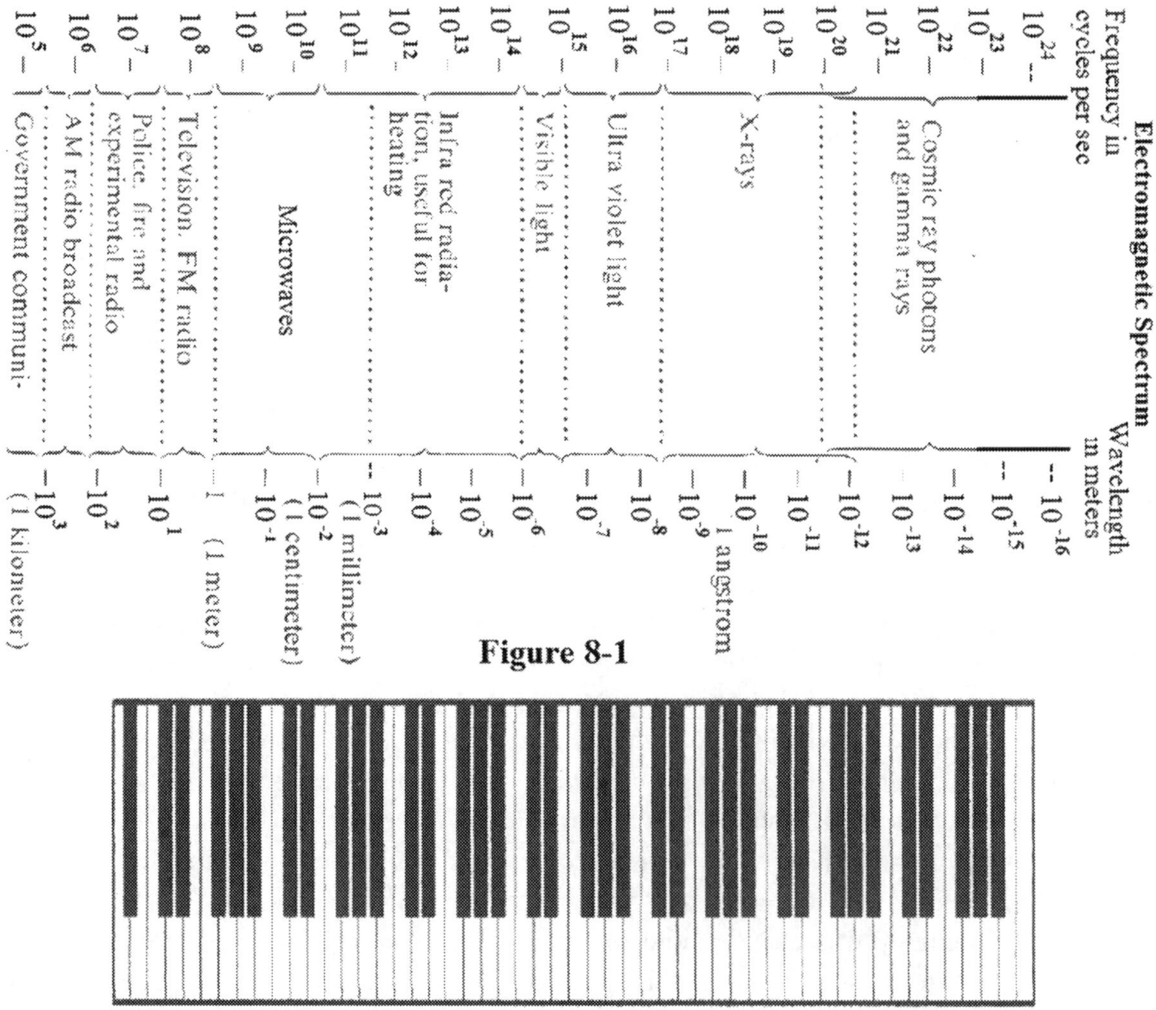

Figure 8-1

For those not familiar with an electromagnetic spectrum, imagine it as being analogous to a piano keyboard displayed above. And think of the different groups of electromagnetic waves as being analogous to the various scales, with the higher pitched notes to the right and the lower pitched notes to the left. But don't confuse the high frequency microwaves, which can travel through empty space in packets called photons at the speed of light, with high pitched sound waves that travel through air in individual notes at a much lower speed. All electromagnetic waves carry energy, and are powerful. They range from cosmic waves down to microwaves, like high C to low A on the piano keyboard. Cooling and stretching changes cosmic rays into gamma rays, followed by X-Rays, which can penetrate right through a body and expose a photograph on the other side. Continuing to the left in the spectrum past UV light that can cause sun burns, the narrow band of visible light is encountered. These electromagnetic waves allow our eyes to see. Farther to the left, we pass infrared waves, followed by microwaves used to cook food, and finely we arrive at electromagnetic waves used for TV and radio communication.

More Ideas Probed for the Origin of the Universe

One of the postulates of Einstein's special theory of relativity, is that inside any inertial frame of reference all laws of physics will be obeyed in every arena, as in mechanical, thermal, optical or electrical etc. These laws will obtain inside two or more inertial reference frames that are at rest or in motion at constant velocity with respect to each other, or if one passes while another one is moving slower. However, a difference in the amount of time measured for the same event by observers located in different reference frames can occur. It is known as time dilation.

Examples of Two Inertial Reference Frames

Two different reference frames A' and B' are represented below. B' is stationary on Earth and A' is moving in a semitrailer with velocity v. A' sees the pulse of light move to the mirror and back in the time it takes for the light to travel (2X). B' sees the pulse of light take a longer path, directed by A' changing the angle of the flash gun so its light pulse will strike the moving mirror, and take more time than (2X) to complete. The difference in timing of the same event as measured between two reference frames is known as time dilation. The speed of light is constant, but takes more time to travel farther.

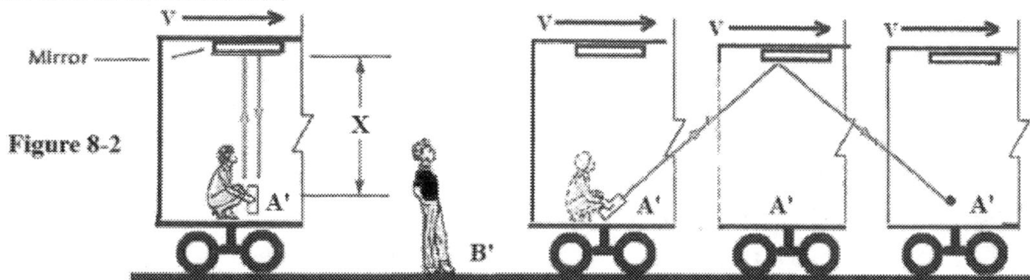

Adapted from College Physics, Tenth edition Serway and Vuille, page 902.

The reference frame of one Earth rotation on its axis measures the rate of <u>Earth's time flow</u>, in days quite accurately. Man has divided each day's time (one rotation) into hours, minutes and seconds. In contrast to this accuracy, a cosmic clock started keeping cosmic time by counting electromagnetic waves, beginning just after the birth of the universe in the cosmic reference frame of the expanding primordial universe. Though every wavelength represents a tick of the cosmic clock, each one in succession measures a slight decrease in the rate of cosmic time flow, due to the red shift stretching of each wave caused by the expansion of the universe. The progressive increase in the length of each succeeding wave, means that fewer waves can pass a given point while traveling at the constant speed of light. In other words, the number of ticks per unit of time decreases, which causes the cosmic clock to slow down. Just as one rotation of Earth has been divided into hours, minutes, and seconds, so one wavelength is measured in man-made meters, and the time it takes for one wave to pass a given point is measured in cycles of man-made seconds or cps. Since the beginning, this process has been ongoing in the leftover electromagnetic radiation that Penzias and Wilson discovered, but don't forget, the cosmic clock began keeping cosmic time long before the Earth's rotations began.

One wavelength can be measured from any place on one wave to the same place on the next wave as the Greek letter, lambda, delineates in the diagram below.

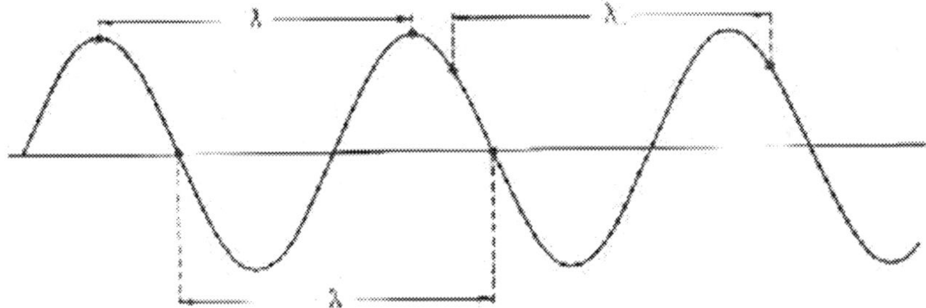

At the outset, very short electromagnetic wavelengths would have measured the rate of cosmic time flow very rapidly, when compared to the rate of Earth's time flow, measured in one Earth rotation. However, as the universe expanded and cooled, the decreasing rate of cosmic time flow measured by the continual lengthening of the electromagnetic waves in its reference frame, moved ever closer to the constant rate of Earth's time flow in its frame of referance.[18] But keep in mind that as electromagnetic waves move through space at the constant speed of light, each one continues to lengthen, thus slowing the rate of cosmic time flow while the rate of time flow of one Earth rotation remains constant.

Early Speculation Regarding the Formation of Mass and Passing of Cosmic Time The production of the mass for the entire universe can be understood in terms of physics as following Einstein's famous equation, $E = mc^2$, the back bone, so to speak, of <u>special relativity</u>. His equation explains that energy can be converted into mass, or that mass can revert back into energy, as when an atomic bomb explodes. However, all of the mass for the entire universe was produced in a very short amount of cosmic time, after the sudden release of an infinite amount of electromagnetic energy in an infinitely small space, just after zero time.[19] Though some of this gigantic amount of mass would need modification later, it was large enough in size, at the outset, to provide all the physical material needed, to make the billions of stars, planets, and moons etc. for the billions of galaxies in the entire universe. In other words, it was a one-time event.

This sudden huge blast of hot radiant energy, was nicknamed the Big Bang, from a snide remark made by Fred Hoyle, an English physicist.[20] Nevertheless, this is how the birth of the universe got its name. But one question remained, when had this happened?

The temperature of the universe is measured in Kelvin or K degrees not centigrade. However, the markings on a Kelvin thermometer are the same as those on the centigrade, but the zero on the Kelvin is absolute zero, which is 273.15K degrees below the freezing point of water.[21]

If the Kelvin temperature could be known at the instant that stable mass formed, and if the instantaneous temperature of the universe could be known now, all the amount of cosmic time that passed between these two event markers could be calculated, by analyzing the radiation of the Cosmic Microwave Background or CMB.

Three questions, beg to be answered. What was the instantaneous temperature of the universe when stable mass formed very close to time zero just after the Big Bang, and what is the instantaneous temperature of the universe now? Also, what is the rate of reduction in cosmic time flow as it decreases. The first was obtained, indirectly, when nuclear physicists found it took energy equivalent to a temperature of $1x10^{13}$K to smash an atom into very small bits of unstable matter called quarks. Therefore in the reverse, during the creation of atoms, the unstable quarks could become stable when confined inside protons or neutrons, at a temperature of $1x10^{13}$K. This is known as the temperature of quark confinement[22]. The present temperature of the universe was measured by the WMAP Satellite and found to be 2.725K.[23] As per Wiens displacement law, the ratio of reduction in the rate of cosmic time flow is proportional to the decrease in temperature from $1x10^{13}$K down to 2.725K.

"When the universe doubles in size, its temperature falls by half." So says Stephen Hawking in his book, *A Brief History of Time* on page 116. Armed with this information and more, we can know that every time the universe doubles in size the temperature of the universe decreases by half and the rate of cosmic time flow also drops by half, allowing it to approach closer and closer to the rate of Earth time.

The equation $N=N_0 e^{-kt}$ can be used to find the time it will take for half of any exponentially decreasing entity to be reached. The small letter k represents the rate of change in the temperatures between 2.725K and $1x10^{13}$K, which is proportional to the decreasing rate of cosmic time flow occurring between quark confinement and that of the universe now. These two temperatures form a fraction of 2.725K/$1x10^{13}$K, which when reduced, equals 1/$3.669x10^{12}$. N_0 equals $1x10^{13}$K, the instantaneous temperature of the universe at quark confinement. N stands for the instantaneous temperature of $5x10^{12}$K, the temperature of the universe at the time of its first doubling. The t equals cosmic time, measured in cosmic days. Since all of these numbers are derived from the cosmos, they occur in the cosmic reference frame. If the present cosmic temperature of the universe of 2.725K is represented proportionally to the time of one Earth rotation and used to solve the above equation, half of the number of initial rapidly flowing cosmic days will be found to fit snugly into one of our present 24 hour days.

Finding the Amount of the First Half of Cosmic Time Using Equation $N=N_0\ e^{-kt}$

1. $5x10^{12} = 1x10^{13}e^{-t\ x\ 1/3.669x10\wedge 12}$
2. $5x10^{12}/1x10^{13} = e^{-t\ x\ 1/3.669x10\wedge 12}$
3. $0.5 = e^{-t\ x\ 1/3.669x10\wedge 12}$
4. $Ln\ 0.5 = ln\ e^{-t\ x\ 1/3.669x10\wedge 12}$

Using the property of log rhythms, ln $e^x = x$, then ln $e^{-t\ 1/3.669x10\wedge 12}$, equals - t x $1/3.669x10^{12}$.

5. Then the entire equation becomes -0.6931 = - t x $1/3.669x10^{12}$
6. t = 0.6931 x $3.669x10^{12}$= $2.5429839x10^{12}$, the number of cosmic days per Earth day.
7. When $2.5429839x10^{12}$ cosmic days are divided by $3.6525x10^2$, the number of days in one Earth year, this yields 6.962 billion cosmic years that passed during the first Earth day.

The above calculations represent the first doubling in size of the universe, which caused its temperature to drop by half and its electromagnetic wavelengths to double. The 6.962 billion years of cosmic time that passed in the cosmic reference frame during the first doubling, corresponds with the first 24 hours of Earth time that passed in its reference frame, described in Genesis One. From God's perspective in His frame of reference, it took 6.962 billion years of rapidly flowing cosmic time for Him to complete all the creative acts needed to finish Creation Day One. But man looking back from his reference frame at the same event measured the passage of only one 24 hour day. This difference in timing of the same event by observers located in two reference frames is called time dilation.

With each additional doubling in size of the universe, the rate of cosmic time flow is halved. Therefore, the amount of cosmic time that passes during each additional Genesis Day, is obtained by dividing the length of the previous Genesis Day by two.

1. During 24 hours of Genesis Day One, 6.962 billion cosmic years passed.
2. During 24 hours of Genesis Day Two, 3.481 billion cosmic years passed.
3. During 24 hours of Genesis Day Three, 1.740 billion cosmic years passed.
4. During 24 hours of Genesis Day Four, 0.870 billion cosmic years passed.
5. During 24 hours of Genesis Day Five, 0.435 billion cosmic years passed.
6. During 24 hours of Genesis Day Six, 0.217 billion cosmic years passed.

Total 13.705 billion cosmic years

It took 13.705 billion cosmic years of the very rapidly flowing, though continuously decreasing rate of cosmic time flow to complete creation, but it took only six 24 hr. Earth days for the same event to occur. This shows the huge differences in measurements of time that can be observed for the same event by observers in two different reference frames. This is an example Time dilation. But, it now known that the rate at which the universe increases in size is accelerating[24]. If so, this will cause the rate of cosmic time to decrease faster than previously figured.

During the creation of the universe, the Bible describes six basically unvarying 24-hour days of Earth time pass by Earth's reference frame. When the steady rate of all six 24 hour time segments of Earth days is converted into the extremely rapidly flowing, though continuously decreasing, rate of cosmic time, the creation week used a total of 13.705 billion cosmic years. The Planck Space Craft Satellite measured the age of the universe to be 13.798 billion years from the Big Bang to the present.[25] But from God's cosmic reference frame these would be cosmic years, which would include creation time, plus the cosmic time that has passed since.

The changing ratios between the fixed time flow of one Earth day and the continuous halving of the rate of cosmic time flow, with each doubling in size of the universe, may be likened to buying an ounce of gold with Mexican pesos on six successive days. If the Mexican Government doubles the value of the peso each day for 6 days, half as many pesos will be needed to buy gold (fixed at $1250/ounce) on each remaining day. This arithmetic works when applied to the fixed rate of Earth time flow and the rate of cosmic time flow halved as the universe doubles in size.

In the cosmic reference frame, the Big Bang created radioactive isotopes with half-lives acting like tiny clocks; preserved in rocks in the Earth's reference frame. According to the equation $E=mc^2$, every atom in the entire universe, including each one in radioactive isotopes was made during the Big Bang from energy of electromagnetic radiation. This placed a relatively large amount of atomic energy inside each atom, including those of all radioactive isotopes. But the energy inside an atom of every radioactive isotope is unstable and seeks stability by expelling from itself, one of three kinds of energy segments. The release of one of these segments usually changes the unstable radioactive isotopic atom into a different, more stable element in the periodic table. The time that it takes for half of a given primordial radioactive element to decay into a stable one, is its half-life, measured in years of cosmic time.

Every kind of radioactive isotope has its own unique unvarying length of half-life time, and each has been accurately measured. Those with a short half-life quickly decayed away. But those with long half-lives can still be found existing ever since the Big Bang, confined in Earth rocks. In them, their half-lives are slowly ticking away time. When read from present to past, these clocks can yield the approximate age of the rocks in cosmic time.

The balance of this chapter deals with the Biblical creation of Earthly life; therefore, the origin of life from evolution's point of view needs to be examined. It looks backward in time, from the present to the past, whereas creation looks forward from the past to the present. Paleontologists use the long half-life decay rates of various radioactive isotopes found in rocks to measure the amount of cosmic time from present to past that fossils have been entombed in them.

From these findings, a calendar of evolution events has emerged. Single-celled prokaryotic life supposedly arose spontaneously about 3.8 billion cosmic years ago. It took about 2 billion more cosmic years for these prokaryotes to evolve into single-celled eukaryotes, and about another billion additional cosmic years to evolve into multi-celled eukaryotes. Later, during the Cambrian explosion, about 20 new phyla of multi-celled eukaryotes arose rather quickly. This calendar of postulated evolutionary events is displayed on the top of the next page.

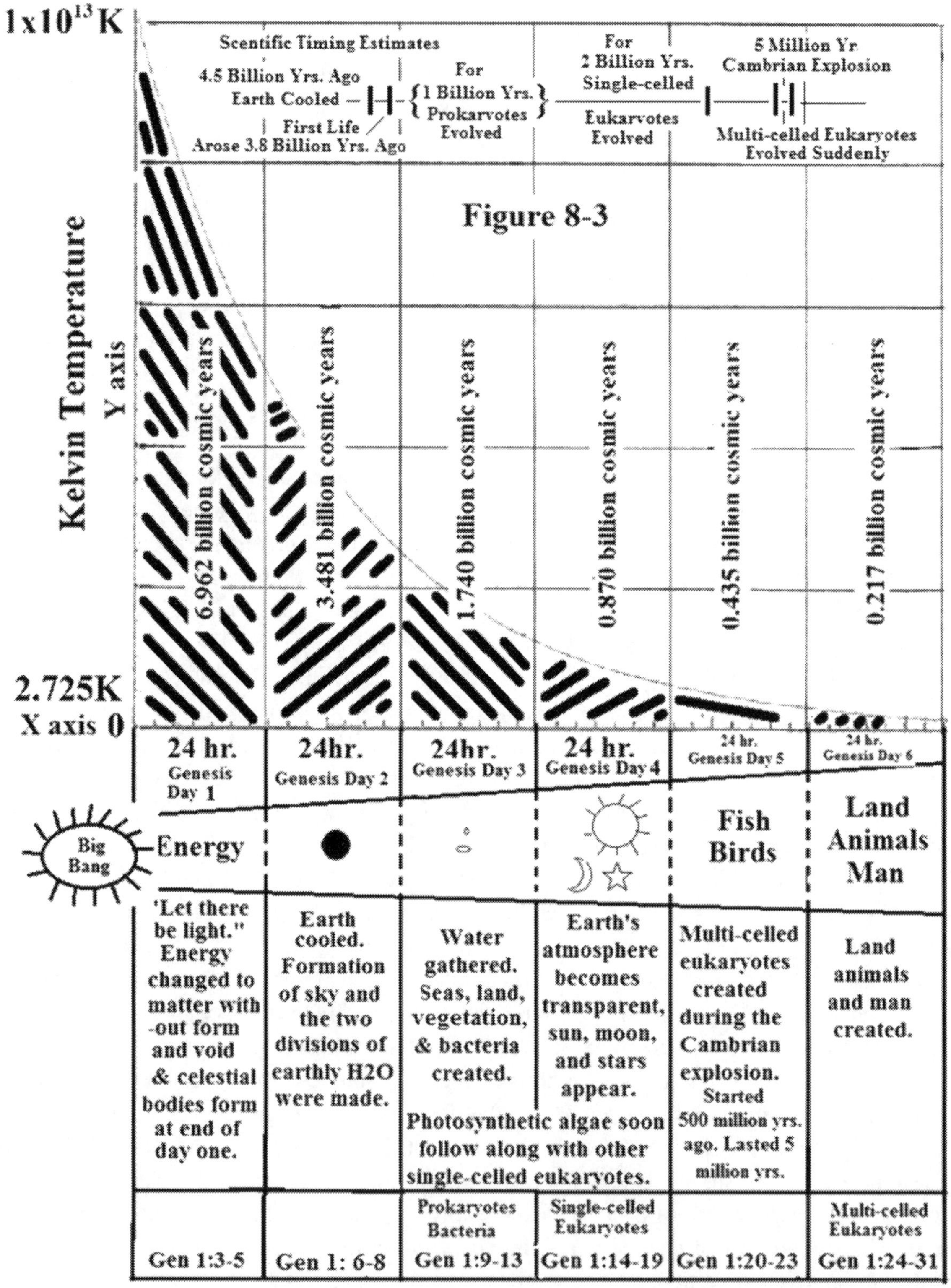

1x10[13]K

Scentific Timing Estimates

4.5 Billion Yrs. Ago Earth Cooled

First Life Arose 3.8 Billion Yrs. Ago

For {1 Billion Yrs. Prokaryotes} Evolved

For 2 Billion Yrs. Single-celled Eukaryotes Evolved

5 Million Yr Cambrian Explosion

Multi-celled Eukaryotes Evolved Suddenly

Figure 8-3

Kelvin Temperature

Y axis

6.962 billion cosmic years

3.481 billion cosmic years

1.740 billion cosmic years

0.870 billion cosmic years

0.435 billion cosmic years

0.217 billion cosmic years

2.725K

X axis 0

24 hr. Genesis Day 1	24hr. Genesis Day 2	24hr. Genesis Day 3	24 hr. Genesis Day 4	24 hr. Genesis Day 5	24 hr. Genesis Day 6
Big Bang — Energy				Fish Birds	Land Animals Man
'Let there be light." Energy changed to matter with -out form and void & celestial bodies form at end of day one.	Earth cooled. Formation of sky and the two divisions of earthly H_2O were made.	Water gathered. Seas, land, vegetation, & bacteria created.	Earth's atmosphere becomes transparent, sun, moon, and stars appear.	Multi-celled eukaryotes created during the Cambrian explosion. Started 500 million yrs. ago. Lasted 5 million yrs.	Land animals and man created.
		Photosynthetic algae soon follow along with other single-celled eukaryotes.			
		Prokaryotes Bacteria	Single-celled Eukaryotes		Multi-celled Eukaryotes
Gen 1:3-5	Gen 1: 6-8	Gen 1:9-13	Gen 1:14-19	Gen 1:20-23	Gen 1:24-31

Look at the graph-diagram on the previous page. There, the sequence of creation events described in Genesis 1 are coordinated with the estimated timing of supposed evolutionary events. The former occurred in 24 hour Earth days, shown below the curve in the middle of the page, whereas the latter are shown in the graph-diagram in years of cosmic time at the top. Notice how close the estimated timing of the postulated evolutionary events in cosmic time are matched by creation events displayed in Earth days for the same occurrences. When the timing and sequence of these events are all superimposed on top of each other, as in the graph-diagram, the cosmic timing of most evolutionary events coincides very closely to the Biblical accounts of these happenings in Earth time.

The graph-diagram very succinctly, delineates the six days of creation described in Genesis chapter one. The number of years of cosmic time that passed during each 24 hour Genesis Day is shown with diagonal lines in the mid-portion of the graph-diagram. Notice that the area under the exponential curve representing each Genesis day in cosmic time, becomes halved with each doubling in size of the universe.

Carefully compare the Biblical descriptions of each of the six individual Creation Days shown underneath the graph-diagram with what science portrays as taking place in the universe during the same cosmic time represented above. This shows that immediately after the "Big Bang," displayed with an oval burst in the graph-diagram that the universe was "without form and void," as described at the outset in Genesis. "In the beginning God created the heavens and the earth. And the earth was without form, and void; and darkness was upon the face of the deep. And the Spirit of God moved upon the face of the waters." Genesis 1:1-2 (KJV). These verses form a very brief preview description of the entire scope of God's creative acts in the universe.

Then verses 3-5 jump back, in time, to before the heavens and the earth were created, to Genesis Day One, when God said, "Let there be light." Preceding light, there was only darkness. "God called the light day and the darkness He called night." Light is the only kind of electromagnetic wave with which ancient Hebrews would have been acquainted. The burst of light is representative of the gigantic quantity of high-energy electromagnetic waves, which God released instantaneously "by the breath of his mouth," (Psalm 33:6 KJV) in a space of zero size, containing zero matter, at zero time. After this primordial event, known as the "Big Bang," God condensed energy into matter. Later, this created matter, provided all the material needed to make every celestial body, including our earth. Most likely the creation of these celestial bodies occurred toward the end of Genesis Day One and may have continued into Genesis Day Two.

The description of the Earth's creation continues in verses 6-8, which tells us that God separated the water into two expanses. But as the preview says, darkness existed over the "surface of the deep," so we can conclude that the sun had not yet ignited. Nevertheless, the earth's surface had cooled enough from its molten state so that liquid water could exist. (The Earth's center is still molten as noted by volcanic eruptions.) This portion of the Second Genesis Day's description is a flashback reference to the water first mentioned in the preview of Genesis 1:1-2. On Genesis Day Three, God created seas, dry land, bacteria, photosynthetic algae, vegetation and trees. Notice that photosynthesis could start, but from what light source needed?

The scripture text describing what supposedly was created on the Fourth Genesis Day reads, "And God said, 'Let there be lights in the expanse of the sky ... and let them serve as signs to mark seasons and days and years'" (Genesis 1:14 NIV). These lights, of course, were the sun, moon, and stars. One of their functions was to mark off days on Earth. From this, we can surmise that the sun had already ignited but was not visible on earth even though its diffuse light was. By Genesis Day Four, the vegetation created on Day Three had, via photosynthesis, produced oxygen from carbon dioxide and cleared the atmosphere of "vog," (fog and volcanic smoke) and other atmospheric pollutants left over from the earth's molten state. This allowed the light from the previously existing sun to shine through the cleared atmosphere, as though it had just been created on Genesis Day Four. On creation Day Five, God created fish and birds. Day Six, He used to create all kinds of other animal species and man. Genesis 1:9-31. "And God saw ... all was very good."

Even though the timing of the estimates of evolutionary events seem to concur with the scriptural events recorded in Genesis 1, these findings do not prove evolution wrong or right. They only show that neither is proved contradictory to the other, but only that the timing of each can be logically explained depending in which reference frame the observer is located.

From this point on, whenever the words **"millions or billions of years"** are encountered in this book, let the reader know that the time involved is **cosmic time** and not **earth time**.

SUMMARY

1. Albert Einstein (1879-1955) discovered the relationship between energy, space, mass, and time. It was epitomized by his famous equation $E=mc^2$. Mass, space, and time cannot exist independent of each other, but are preceded by energy. This discovery formed the basis of Special Relativity. When he connected these concepts mathematically with gravity in 1915, (General Theory of Relativity) his equations showed that the universe was curved, expanding, and finite. These concepts meant that the universe had to have had a beginning, because at some point in time in the distant past, all matter had to have existed close together. Einstein did not believe his own equations, so he added a cosmological constant to them, to make them static, because he didn't think the universe could go on expanding for so long.
2. Friedmann found that Einstein had made an algebraic mistake in his equations. With the algebraic mistake removed, along with the cosmological constant, He was probably the first to believe in Einstein's general theory, including Einstein himself. Friedmann was probably the first to understand that the universe was expanding.
3. Following the lead of Slipher; Hubble (during the 1920s) observed through the 100-inch telescope at Mt. Wilson Observatory in California, that the universe was expanding everywhere he looked. These findings caused Einstein to remove the cosmological constant, saying it was the biggest mistake of his life.
4. The first law of thermodynamics states; although energy can be changed from one form into another, it can never be created or destroyed. Therefore, the total amount of energy in the universe has remained constant, since its origin at the Big Bang. The Second Law of Thermodynamics states, energy can move only from a higher state to a lower state, never the reverse. Because each transfer is never 100% efficient, some energy is wasted and can never be recovered again. Energy that is not able to perform work, because it is in too low a state is called entropy. This wasted energy goes to the reservoir of entropy, which is growing bigger with each passing second and energy for doing work is constantly getting smaller. Entropy is always increasing in the universe. The segment of energy available to perform work, plus the unusable energy already in the lake of entropy; when added together, is constant. If the universe was infinitely old, all of the energy in a high enough state to do work; would have already been spent.
5. The trio of Einstein's equations, Slipher and Hubble's discovery of an expanding universe, and increasing entropy each points to a beginning of our universe.
6. The rate of time flow can be different for different reference frames in the universe now and at different times in the past.

Chapter 9

Primordial Soup and Life's Origin

This chapter has to do with the chemical origin of life or chemical evolution. You will see as the concepts of chemical evolution are explored that scientists involved in this endeavor try to understand how life might have arisen spontaneously from self-assembly of chemicals billions of years ago. They call this the bottom up approach. This is in contrast to biological evolution, which Darwin first described in the "Origin of Species." He started with living biota and worked his way down. Hence, this is called the top down approach. Evolutionary theorists hope to unite these two concepts some time in the future. As we start to explore the probabilities of chemical evolution, one important item must be pointed out and kept in mind through the analysis.

You will learn in Chapter 10 that natural selection is thought to be the driving force of biological evolution. It can only choose between living organisms, allowing the more fit to prosper and the less fit to become extinct. From this has come the saying, survival of the fittest. Because non-living molecules existing in an open system cannot compete for survival as do living organisms, natural selection can have no effect on molecular outcomes in the abiotic world. All of the molecules in the proposed primordial soup exist in an open system. This means that none of them have access to controlled energy with the proper intensity and direction to build more complex molecules. These are the attributes of a confining physical system. Therefore, among abiotic molecules there can be no competition between any two of them to determine which is fittest. There is no selecting process in the abiotic environment to choose one molecule over another, so natural selection cannot make choices between non-living molecules.

In a private letter to a friend, Darwin hinted that life forms might have arisen spontaneously in a warm little pond. This is what he wrote; *"It is often said that all of the conditions for the first production of a living organism are now present which could ever have been present. But if (and oh! what a big if!) we could conceive in some warm little pond, with all sorts of ammonia and phosphate salts, light, heat, electricity present, that a protein was formed ready to undergo still more complex changes at the present day such matter would be instantly devoured or absorbed which would not have been the case before living creatures were formed."*[1] Later scientists have called this warm little pond the primordial soup. Even though he postulated that a protein might undergo more complex changes in his little pond, this proposed change must not be confused with mutations. They are confined to living organisms, not abiotic chemicals. However, abiotic chemical reactions bound by the two laws of thermodynamics operating in an open physical system, do undergo change but usually not toward more complexity.

It is possible to conceive of a warm "pond" suggested by Darwin, existing around some deep hot ocean vent or hot geyser near the surface. But would the "pond" contain the correct chemicals each having access to energy at the correct level to direct random construction by pure chance of the proposed protein that Darwin suggested might have arisen spontaneously? Would they be ready to undergo still more complex changes? (The author recognizes that in the mid 1800s, Darwin's use of the word protein did not carry the same understanding of the use of that word today. Notwithstanding, it is certain that Darwin was referring to some type of chemical needed for life to begin, that he postulated would undergo complex changes.) And because any complex carbon molecule needed for life, located in an abiotic environment would be functionless, there would be no mechanism available to drive it to more complexity. For this to occur, it would require a confining physical system with the application of the correct amount of energy to propel a given chemical reaction in the correct direction. Natural selection could not provide the impetus for more change, because its effects are confined to living organisms. Therefore, for life to arise spontaneously, it must rely on total randomness in an open physical system to complete the process. Darwin realized that his proposal had a considerable degree of improbability when he prefaced it with, "But if (and oh! what a big if!)," so let's try to calculate just how big the "**if**" really is. His proposal would require a pond big enough, containing the proper amount and kind of chemicals, and must be in existence long enough for the chemicals to randomly produce the needed result, be they DNA, RNA, or proteins.

To get an idea of the complexity of Darwin's proposed pond, consider the kind of chemicals that would have to be present and how long they would have to exist to produce by pure random chance just one specific protein composed of only 100 amino acids. While affiliated with the Museum of Vertebrate Zoology at University of California at Berkeley, Monroe Strickberger calculated the following; "*By similar reasoning, the chances for most complex organic structures to arise spontaneously are infinitesimally small. Even a small enzymatic sequence of 100 amino acids would have only one chance in 20^{100} (= 10^{130}) to arise randomly, since there are 20 possible kinds of different amino acids for each position in the sequence. Thus, if we randomly generated a new 100-amino-acid-long sequence each second, we could expect such a given enzyme to appear only once in 4 x 10^{122} years!*"[2]

Of course this ignores catalysis of any sort which, in the case of enzymes (proteins) is critical to making chemical reactions of life proceed much more rapidly than simple solution chemistry. But one must ask where did the first protein enzyme come from? Then we are back to nearly infinite improbabilities. Someone might say this also assumes that all 20^{100} proteins will be needed to produce progression to self-organizing systems needed for life itself. However, on closer examination Mr. Strickberger was only calculating the chances for random production of only one specific protein molecule composed of only 100 amino acids. If that protein happened to be an enzyme, the chances for its random production plummets as close to zero as can be imagined in the time available.

Richard Dawkins tried to minimize another big "if" in his book *Climbing Mt. Improbable*, where he compared evolution with climbing a mountain. Dawkins pointed out that most people who do not accept the evolutionary theory think it requires a giant leap up the side of the mountain whose face is a sheer cliff. They find this impossible, and so reject the theory. Dawkins said there was another side to this imaginary mountain that is a gentle slope, and that evolution has ascended this side of the mountain, *"inch by million year inch."*[3] But Dawkins is wrong; there is not just one sheer cliff for evolution to ascend, but three. The first leap is from abiotic to the biotic. The second is from the prokaryote to the single-celled eukaryote and the third is the leap from the single-celled eukaryote to the multi-celled eukaryote.

The first giant leap from the abiotic to the biotic will be analyzed in this chapter. Here is true chemical evolution. We will start from the bottom up, so to speak, looking for evidence for the first giant leap. Stanley Miller's famous 1952 experiment was designed to demonstrate how this leap might have spontaneously occurred. Miller based his experiment on what he believed to be the composition of the Earth's atmosphere 3.5 billion years ago that had been postulated in 1930 by two scientists, Alexandre I. Oparim and J.B.S. Haldane. Each of these men independently had suggested that the early Earth's atmosphere was reducing in nature, lacking oxygen, and was composed mostly of hydrogen, methane, and ammonia (H_2, CH_4 NH_3). Hydrogen is the most abundant element on Earth with two atoms of hydrogen and one atom of oxygen (H_2O) for every molecule of water. However, hydrogen, methane and ammonia are not now present to any significant degree in Earth's atmosphere. Methane, a chemical name for natural gas, was thought to be a major component of the early Earth's atmosphere. It has the chemical formula CH_4. The other gas, ammonia (NH_3), has a very irritating odor and dissolves instantly in water.

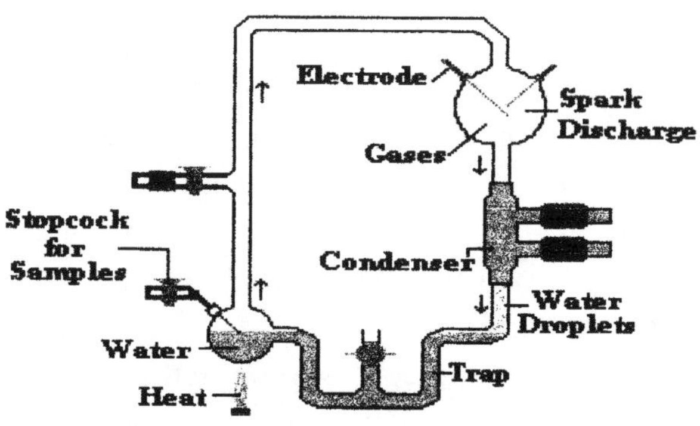

Figure 9-1
Stanley Miller's Experimental Apparatus (a confining physical system)

Miller introduced a mixture of these gases into the apparatus shown above, which you will recognize as a confining physical system. Keep in mind that any properly constructed confining physical system is an engine. For it to work, the intensity of the energy must be correct and the forces must be correctly directed. Along with the gases, he also added water to the bottom flask that he said represented the ocean. He gently heated this so that water vapors would rise, representing evaporation from the ocean's surface caused by the sun's heat. These vapors, along with the gases, passed through an electric spark that simulated lightning. The vapors then passed through a cooling device called a condenser, where the vapors condensed back into liquid form again. This supposedly represented rain. The liquid fell back into the "ocean" flask, and the cycle was repeated. Miller allowed this process to go on continuously for several days. Later, when he analyzed the contents, to his amazement he found that the apparatus contained six amino acids found in living cells, along with many other organic molecules not associated with life. The heat from the flame and the electric sparking device supplied the correct intensity of energy to the inside of Miller's confining physical system (an engine), the walls of which forced the chemical reactions in the desired direction. Both of these energy sources followed the second law of thermodynamics where a larger amount of order was expended to obtain a smaller amount of a different kind of order. This confining system supplied the intensity and direction needed to build the six different kinds of amino acids from the gas molecules originally confined there, as well as other non-biological amino acids. Though not very efficient, it did follow the second law of thermodynamics. To everyone's surprise, these amino acids were similar in type and concentration of those found in a carbonaceous meteorite that fell to Earth in Australia,[4] suggesting to some that Miller's experiment probably mimicked natural processes that occur in outer space.

Other scientists have conducted similar, so-called prebiotic experiments since Miller's seminal one with similar apparatuses (more confining physical systems). These experiments have yielded many other chemicals of life. The conclusion that many evolutionary scientists make is that life-giving chemicals could have arisen on the early Earth spontaneously and later could have come together to produce life. That Miller's experiment produced similar chemicals found in similar concentrations in the Australian Murchison meteorite, furnished even more credence that Miller's experiment was something which actually transpired somewhere in the universe, "*a homemade Garden of Eden*," as Stuart Kauffman called it.[5]

Just because certain amino acids result when hydrogen, methane, ammonia, and water are heated in a confining physical system, and the vapors resulting from this mixture are forced through a chamber containing electric sparks, does not prove life arose this way. Miller assumed that the Earth's early atmosphere had very little oxygen present and that it was composed of all those other gases. This condition may or may not have been the case as later data seem to indicate.[6]

Though his experiment produced six amino acids found in living cells, fourteen more are needed for life, and they must display homochirility. This term means that all organic molecules of a given kind must have the same configuration. They cannot be mirror images of each other, but must either be right-handed or left-handed, similar to either one of your two hands. Each living cell needs thousands and thousands of the 20 different kinds of biological amino acids used to synthesize proteins, and all must be left-handed, except glycine. Each protein has a specific sequence of amino acids in order to form just one protein molecule. Escherichia coli, one of the most common bacteria (prokaryote), has about 4,300 kinds of genes needed for its metabolism. Most of these genes contain patterns for specific proteins. With that in mind, the six biological amino acids that formed randomly in the Miller experiment, (except for glycine) would have been a mixture of half right-handed and half left-handed molecules.[7]

Let's follow this idea of origin of life a lot farther. In the many decades since Miller performed this so-called water-shed experiment, biochemistry has made its greatest strides. Since the early 1950s, the structure of DNA and the genetic code that it contains have been deciphered and understood for the first time.

What has been discovered is that there is complete interdependence of DNA and proteins. There are at least seven proteins, six of which are enzymes needed to replicate DNA. But all information for making each of these proteins is stored in the DNA itself. Of course, RNA acts as a chemical intermediary that takes the message from the DNA to the ribosome for making proteins. The instructions for making the RNA and the ribosome also are stored in the DNA. RNA polymerase, which transcribes the RNA from DNA, is also a protein enzyme whose pattern is stored in the DNA. RNA and proteins cannot exist without DNA and vice versa. Like the chicken or the egg, which came first, the DNA, RNA or the proteins? Leslie Orgel, a prominent biochemist, much of whose professional career has been spent searching for a chemical pathway for the origin of life, points out that it would be extremely improbable to have DNA, RNA, and proteins arise simultaneously in the postulated primordial soup. Today, genetic information flows from DNA to RNA to proteins. To get around this conundrum about the simultaneous origin of DNA and proteins, some scientists have proposed (namely Leslie Orgel, Carl R Woese, and Francis Crick) that RNA came first and that proteins and DNA came along later. This idea seems to have caught on in the scientific community, and has been dubbed the RNA world. However, for DNA and proteins to have evolved from RNA, the RNA had to display two properties at the outset. RNA would have had to act as its own replicating catalyst and as its own repository of genetic information to be a precursor of protein synthesis.[8] With the discovery of ribozymes independently by Thomas Cech and Sidney Altman in the early 1980s, many scientists believed that the replicating catalyst problem had been solved. However, randomly produced RNA would only contain a gibberish of genetic information.

So what chance does an RNA genetic message have of arising spontaneously in the primordial soup? In the prebiotic experiments performed thus far, starting with Miller's and continuing to the present time, all five bases (guanine, cytosine, adenine, thymine, and uracil) have been recovered. However, without the appropriate protein enzymes, workers in this field have had trouble getting ribose to form in adequate quantity and purity. As you will remember, the "R" in RNA stands for the sugar ribose, which is a component of RNA. Without ribose, there could be no RNA appearing in the primordial soup.[9] One could take the position that scientists have not yet found the right pathway to form RNA spontaneously or for that matter DNA or proteins. No RNA molecule has been found that can direct its own replication to produce more RNAs.[10]

Another caveat blocking the progress of the prebiotic experiments has come to light. All have been performed with the assumption that there was no, or extremely little, oxygen present in the prebiotic atmosphere. More recent research seems to indicate that oxygen was quite abundant in the Earth's early atmosphere. Oxygen in any significant concentration would have prevented the prebiotic chemical reactions from progressing toward the spontaneous development of simple life.[11]

Even when the biochemists "cheat" and synthesize RNA in the presence of appropriate protein enzymes, the chemical reaction will not proceed if both right- and left-handed nucleotides are placed in equal mixtures together. Living organisms only produce and use right-handed nucleotides (more homochirility). But in prebiotic conditions, if nucleotides could have formed randomly, equal amounts of right and left would have been present. However, protein enzymes were artificially added while carrying out these experiments.[12] (If RNA was supposed to act as its own replicating catalyst and the repository of genetic information that was supposed to have preceded proteins, the enzymes, which are themselves proteins, should not have been allowed in this experiment).

Let's suppose that RNA could have arisen spontaneously in the prebiotic soup and that it had all the characteristics needed for self-replication and for acting as a repository of genetic information. Protein synthesis would have been stymied for two reasons. First, there would have been no ribosomes to manufacture the proteins and, second, the amino acids produced under prebiotic conditions by random means would form in equal numbers of right-handed and left-handed configurations. Only left-handed amino acids (except for glycine) can be used. The codon triplets, all formed from right-handed nucleotides, only choose left-handed amino acids in biological protein construction.

Now let's further tie this into our analysis of the origin of life. The Escherichia coli bacterium has only one circular chromosome composed of 4,391 genes, which are the specifications for about the same number of proteins, less tRNAs and rRNAs. Researchers at the University of Wisconsin at Madison have discovered that there are 4,639,221 base pairs in the genome of E. coli K12 MG1655. Because bacteria have very few, if any, introns (non-translated codons sometimes called junk DNA between genes in their chromosomes), we can divide the number of base pairs by the number of genes to get an approximate average number of base pairs in one gene. This comes out to be 1,056. Because the number of base pairs in the DNA would be the same as the number of individual nucleotides in a single-stranded RNA, then if we divide 1,056 by 3 (the number of nucleotides in one codon), this will give the approximate average number of codons in one gene. Because there would be 352 codons in the average mRNA representing a gene of the above-mentioned E. coli and because each codon specifies one amino acid, this calculation will show that the average E. coli protein would contain approximately 352 amino acids. By pretending that RNA came first in the evolutionary scenario as many evolutionary scientists propose, then the odds of finding an RNA molecule 352 codons long can be estimated.

To make our calculations easier, let's round off the number of triplet codons in one mRNA molecule to 333. This means that there would be an average of 999, or approximately 1,000, nucleotides in each gene. Now, let's consider what the chances would be to randomly construct just one RNA molecule coding for just one biologically active protein, about 333 amino acids long, in the primordial soup. This one RNA would have to be one of many needed to construct the first proto-prokaryote. It requires one codon to specify each amino acid. Three nucleotides compose each codon. So it follows that for a 333 amino acid protein to be properly coded, 999 nucleotides would be needed to construct a specific RNA molecule. Besides the sequences for each codon, we must remember that the three nucleotides composing each codon require specific sequences of nucleotides to designate its code. These nucleotides are randomly united in the supposed reservoir in the prebiotic soup. In this mixture, there had to be eight kinds of nucleotides from which they could be randomly united to form an RNA molecule 999 nucleotides long. This would require at least 8^{999} trials to get one specific RNA molecule.

A question immediately comes to mind. Why were eight nucleotides needed in RNA construction if there are four: adenine, guanine, cytosine, and uracil? If that is true, then why not limit the total to four (instead of eight) that would be used in a biologically active RNA molecule. The reason is that half of the randomly formed nucleotides, if produced in the prebiotic soup, would be right-handed and half left-handed. The importance of homochirility applies here. This would give two of each kind of nucleotide. The number eight was chosen to represent the different kinds of nucleotides that would theoretically form randomly, four right-handed and four left-handed. Therefore, at least eight kinds of nucleotides would be required. The fewest number of different random trials (assuming there are no repeats) needed to guarantee the formation of one specific molecule of mRNA, 999 nucleotides long would be 8 raised to the 999 power (8^{999}). Because evolutionary theory has no memory, goal, or intelligent guidance, this number could be many orders of magnitude higher because without memory any given trial could be repeated an infinite number of times.

Perhaps the random construction would have been lucky and produced a self-replicating RNA molecule on the first or the second trial. Even so, when we consider that with E. coli, this lucky process would have to repeat itself 4,391 times, once for each of the 4,391 genes coded in its genome, it becomes apparent that we are back again to astronomical improbabilities. This is only a nice way of saying that it is impossible. Remember, there would have been no ribosomes if RNA preceded everything else. At some point, the RNA would have to be transcribed in a retro-fashion to form DNA, which would require eight more kinds of nucleotides in the primordial soup. It doesn't take a rocket scientist to see that luck would not repeat itself 4,391 times. You have learned in this chapter that natural selection cannot preferentially select various molecules in abiotic environments because in this situation all biological molecules are functionless and cannot compete with each other for survival as do living organisms.

Eight to the 999th power (8^{999}) changed into scientific notation is 1.538×10^{902}. That would be the same as 1538 with 899 zeros after it. This means that there would be only one chance in 1.538×10^{902} of randomly constructing one mRNA molecule with the correct sequence, assuming no repeat trials of any wrong combinations. However, as already pointed out, with no memory or goal there could be an infinite number of wrong trials. Let's suppose that a method was present, that every second could sort one hundred billion (100,000,000,000) combinations of nucleotides, each 999 long, in this mixture of chemicals. Obviously this is hypothetical, because there was nothing in the primordial soup to do the sorting. There are 31,557,600 seconds in one year. This number multiplied by 100 billion would yield 3.15576 X 10^{18} trials each year. If we divide 1.538 X 10^{902} by 3.156 X 10^{18}, we get 4.87 X 10^{883}. This equals the number of years it would take to guarantee the random construction of one correct mRNA molecule, 999 nucleotides long, from the prebiotic soup if it took all 8^{999} (1.538 X 10^{902}) trials, assuming no repeats. Even if this number were divided in half, which is 2.435 X 10^{883} years (which equals the average time needed to guarantee the formation of all the correctly formed mRNA molecules, again assuming no repeats), it would be many orders of magnitude longer than the time allotted for life to have evolved by random means after the Earth cooled down enough from about 4.5 to 3.8 billion years ago.

Even if the RNA scenario just described is not the correct one, maybe DNA or proteins came first. Even so, the same astronomical improbabilities would exist with their proposed random construction in an open system. As Stephen Hawking pointed out there are many orders of magnitude of wrong combinations, but only one correct one as with shaking up many pieces of a jigsaw puzzle in a box to see if the pieces come together perfectly. If this is true for the random construction of either one of these molecules, as with many pieces of a jigsaw puzzle confined in a box, imagine the huge improbabilities of constructing not just one RNA, DNA, or protein, but all three simultaneously. Having said this, it is common knowledge that as science progresses, new discoveries are made. Perhaps one will come forth in the future, which will prove a nonrandom way for very complicated biological molecules to self construct. However until that time, we are stuck with the extreme unlikelihood described in the above scenario or one similar to it.

Even if the proto-bacterium had only 100 hypothetical RNA genes instead of the 4,391 actual DNA genes that form one E. coli bacterium, an astronomical improbability would still exist to randomly form 99 more hypothetical mRNA genes for the first living organism. Later, this hypothetical organism would require the mRNA to produce DNA along with hundreds of other component parts such as DNA polymerase and ribosomes, all wrapped up in a double-walled lipid membrane.

There is no evidence that Darwin's proposed warm little pond containing ammonia, phosphate salts, light, heat, and electricity ever existed in the prebiotic world. If it ever did exist, the requirements of biochemicals needed for life would have exceeded those in Darwin's little list by many orders of magnitude for his proposed protein compound to have spontaneously made its appearance.

Does it seem reasonable that someone could win 100 or more lotteries on the same day? Origin of life theory is like a chemical lottery that would have us believe that at least 100 complex biochemical molecules came together in one place at the same time to produce the theoretical proto-bacteria. (The number 100 is very conservative. It probably should be in excess of several thousands.) This notion is even more preposterous than winning just 100 monetary lotteries simultaneously. In real monetary lotteries, there is an actual pool of money that always exceeds the amount from which to draw the winnings. There is also a method of choosing who the winner will be, how much will be paid, and the time it will be paid. In the prebiotic realm the evidence for even a small pool containing a concentration of nucleotides is lacking. There also would be no method to select a winner from the proposed theoretical chemical chaos in this hypothetical scene. Because of these constraints, all 100 winners of our proposed chemical evolutionary lotteries would have to have been selected simultaneously from a theoretical pool that didn't exist and by a method that could make no choices without memory, goal, or intelligent guidance. This only shows that something that couldn't happen didn't.

There is another restriction usually not mentioned by evolutionary biologists about the origin of life. That is time. It is known that life appeared relatively soon, geologically speaking, after the Earth cooled enough for life to exist. So there were not billions of years available for random chemical reactions to produce the first living cell. We know that bacteria, though simpler than other life forms, are very complex. Bacteria now appear to be the same, or very similar, to fossils that appear to be 3.5 billion years old. They occupy more ecological niches than any other life form. In fact, the late Stephen Gould calls this the age of bacteria.[13] We have learned that, even though bacteria have only one circular chromosome, an E. coli has a few more than 4,300 genes, specifying many proteins, for its existence. The chances for 4,300 genes composed of as many as 1,000 or more base pairs to have happened by pure random chance is so preposterous it is beyond imagination. However, brilliant men have become so enamored with the origin of life paradigm that it has taken priority over reality and common sense.

The prebiotic experiments of Miller and those that followed demonstrate what smart individuals can cause to happen in small confining physical systems. The flask that represented the ocean in Miller's experiment allowed the chemicals to concentrate there. Even if these gases (bombarded by real lightning or intense UV rays, in the Earth's early atmosphere) could have formed the same amino acids found in Miller's prebiotic experiment, they could not concentrate when falling into the real ocean because the first wave would have eternally separated them by pure dilution alone.

The list of attendees to the twelfth International Conference on the Origin of Life at the University of California at San Diego on July 11-16, 1999, showed that Miller and Leslie Orgel attended. The conclusion of one abstract presented on "Prebiotic Synthesis of Nucleotides" by Geoffrey L. Zubay, from p. 39, cB 1.5 Columbia University, reads as follows: "*Despite the gains that have been made there are many problems that remain unsolved. The linkage of ribose to the purine base is one of the biggest and this is where we are concentrating our effort at the present time.*"

The two purine bases are adenine and guanine. Zubay says that he can't get adenine and guanine to unite with ribose under postulated prebiotic conditions. So you can see ribose and the formation of two of the nucleotides found in RNA are still two of the biggest problems for the origin of life scenario.

The calculated weight of a single bacterium composed of 100 billion atoms, weighs in at less than 10^{-12} grams.[14] If the biochemists cannot get the more basically needed biochemicals of life to come together in their artificially small confined physical systems, how can they expect billions of more atoms to come together randomly in a large open system? With failure of chemical evolution, Darwin's biological evolution is dead before it even starts.

ENTROPY

Principles involving the first and second laws of thermodynamics were developed in Chapter 8. These will be reviewed briefly again. The first law states that, although energy can be changed from one form into another, it can never be created or destroyed. Therefore, the total amount of energy present in the universe has remained constant since the Big Bang when all energy came into existence. The second law states that energy can be transformed only from a higher state into a lower state, never in the reverse, and with each transfer some energy is lost. In other words, no transfer is 100 percent efficient. That portion of energy wasted in every transfer is gone forever for use in the production of useful work or the production of a smaller amount of order. It can never be recovered to be used for work. This wasted portion of energy is called entropy. The reservoir of entropy is always increasing, never decreasing in a closed system. The universe as a whole, acts similar to a closed physical system. Entropy is the measure of unusable energy and also a measure of disorder in the universe.

The ultimate result of the second law is also a description of going from order to disorder. When energy is released, it always produces a force, which always has two components, intensity and direction. It is important to note that there is a way to restore order from disorder without breaking the second law of thermodynamics. This can be accomplished only when much more energy is released with a force that has the correct intensity and direction. This always results in the production of a larger portion of entropy and a smaller portion of useful work in the formation of a small amount of order. This is in obedience to the second law. But remember, when any order is brought out of disorder, only directed forces, with the right intensity, can do the job. These requirements can be met in a properly constructed confining physical system. Keep in mind, when any force is released from any energy source, more energy goes to the lake of entropy than is used for the restoration of order from disorder.

When a force is released, as in an accidental factory explosion, the intense force is distributed in an almost infinite number of directions with the factory being severely damaged. Order turns to disorder with nearly 100 percent of the energy released during the explosion going to entropy. However, a very small fraction may have produced some order out of the gross amount of disorder. If, for instance, at the time of the blast a small amount of sand happened to be located inside the factory, the intense heat released could have turned the sand into a small blob of unusable glass. This glass, though unusable, would contain a very small amount of order as compared with the sand from which it was made. Nearly 100 percent of the energy in the accidental factory explosion would be wasted, going to entropy.

These principles show that the ultimate outcome of many energy transfers can produce a small amount of work and larger amount of entropy. This is true for all forms of energy transfer, be it heat, mechanical, electrical, chemical, or atomic. They simultaneously also illustrate the two components of force, namely intensity and direction. Any release of energy, including biological energy transfers, always includes these same components. Energy released from biological chemical reactions, though quite efficient for producing useful work, also looses some energy to entropy. Biological chemical reactions that use energy to force their completion must also display proper intensity and direction. Biological chemical reactions are almost always very heat sensitive, requiring perfect temperature control within the narrow limits required of their respective confining physical systems.

The theory of origin of life requires movement from the disorder of the proposed hypothetical primordial soup to the extreme order of the first life, such as a bacterium. To build the first bacterium from scratch obviously would require millions of individual steps to put together the DNA containing several million base pairs, RNA, and ribosome to manufacture the many proteins, all enclosed inside a double lipid wall. Each step would require the release of just the right amount of energy (intensity) to force the uphill process of construction (direction). Theoretically, it might be conceivable that the very first step could happen in an abiotic environment and could have constructed a few molecules (amino acids) needed for life. Similar to the blob of glass formed in the factory blast, these molecules could have been produced by energy releasing a force that randomly happened to have the right intensity and direction for their production. To increase the chances to produce the proposed molecules, the force with random directions and with uncontrolled intensity must be released in a fairly dense concentration of simple carbon molecules with at least a smattering of simple nitrogen molecules and a few other elements. The existence of primordial soup with these parameters has never been substantiated. These conditions require a confining physical system similar to Miller's prebiotic experiment, where the combination of the two directed forces of heat energy and electrical energy could work on the chemical building blocks over and over again. This also would allow the semi-randomly produced biological molecules to concentrate in the heated flask. However, in the postulated large open system of the proposed, hypothetical primordial soup only random undirected forces with unpredictable intensities would be available to produce the molecules in the first step. Even if a few basic building blocks (chemicals of life) could have formed, dilution would have separated them forever, before they could polymerize into chains of RNA, DNA, or proteins.

The First Overhanging Cliff
Abiotic World to the Biotic

1. The second law of thermodynamics prevents order from arising from disorder without the expenditure of more energy whose force has the proper intensity and direction. There was adequate energy available in the primordial soup, but nothing to give direction. Direction implies a confining physical system. The primordial soup has no such system.

2. Function of complex systems composed of many complex parts cannot begin until all of the parts are positioned in their proper places simultaneously. Then function can begin. The simplest life form, composed of thousands of parts, cannot live until all of its parts are simultaneously assembled. Evolution, controlled by randomness, cannot do this without infinite time available.

3. Homochirility is absolutely necessary for life. This is especially true of the amino acids and their nucleotides. The former are levo and the latter are dextro in their respective configurations. No way has been found for random methods to segregate these isomers.

4. Ribose in significant amounts has not been produced in prebiotic experiments. Without ribose there can be no RNA.

5. In prebiotic experiments, no way has yet been discovered for ribose to link to the purine bases of adenine and guanine.

6. Most prebiotic chemical reactions must be performed in the absence of oxygen. It is now thought that oxygen was present in the abiotic atmosphere.

7. Even if functionless parts could have been produced in an abiotic pond, dilution would prohibit them from coming together.

8. Life appeared soon (geologically speaking) after the earth cooled enough for life to exist. Therefore, there were not billions and billions of years for the accidental random construction of the first life. In addition, Darwin's proposed warm little pond would have been too small for prebiotic chemical reactions to randomly produce the many organic molecules necessary for life to begin.

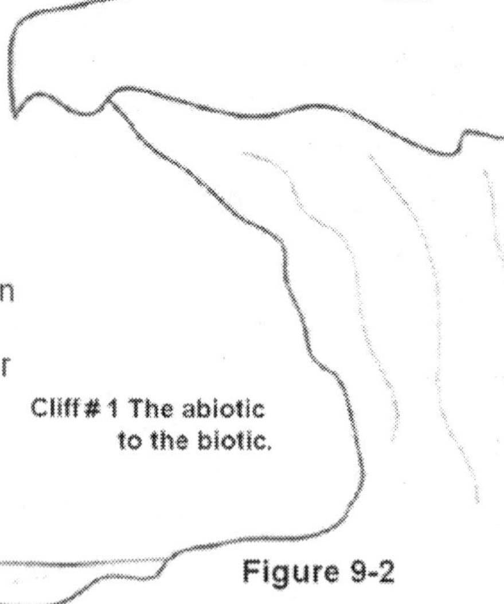

Figure 9-2

Let's suppose that a few amino acids, or other biological molecules, could have been put together by a flash of lightning in the first step, similar to the blob of unusable glass in the factory explosion. The energy releasing the force (lightning) most likely would have been as intense as the factory blast. However, like the explosion, there would have been no specific direction to the force released as there would be in a confining physical system. From a biological point of view, these hypothetically produced biological molecules by themselves would have been functionless. There would be no means to select them from the morass of the proposed primordial soup so there would be no place for them to concentrate in a large open system. Therefore, unlike the confining flask in Stanley Miller's prebiotic experiment, there would be no method for preserving or concentrating the first biological molecules produced by the proposed randomly directed forces with uncontrolled intensities. In addition, there is no guarantee that forces with unpredictable intensities released with random directions, by energy needed in the second step, would not destroy the few needed molecules produced by the first step. Also, unlike Miller's experiment, there is no method to circulate the chemicals around and around again and again to increase the chances for their production. Therefore, each succeeding step would have only one trial. Without some sort of preservation and concentrating mechanism in place (as in a confining physical system) to preserve these hypothetical molecules needed in each succeeding step, the progress of the origin of life becomes hopeless. In accordance with the second law of thermodynamics, only directed force with the proper intensity can produce order from disorder. Force, without controlled intensity and direction, only breeds more disorder. Every living organism is an example of a confining physical system, which is actually a miniature engine, where energy derived from outside itself is expended inside with the proper intensity and direction.

Life cannot exist outside these parameters. It is the lack of controlled energy, releasing forces with the proper intensity and direction in the hypothetical primordial soup, or any other scenario which the origin of life would require, that prohibits their advancement and stops the chemical origin of life in its tracks. This illustrates why the theory of the origin of life supposedly propelled by pure undirected random chemical reactions could not have produced even a very simple life form. Without control of intensity or direction, the laws of heat and thermodynamics essentially prohibit assemblage of even a so-called simple life form. You will see in the next chapter that without the development of the first life, Darwin's Biological Evolutionary Theory was dead before it ever began (Figure 9-2). This is another boundary which confines Darwin's theory to unreality.

SUMMARY

1. Evolution has three cliffs that it must climb.
    a. A simple life forming spontaneously in the abiotic primordial soup is the first cliff.
    b. The evolution of prokaryotes to single-celled eukaryotes is the second cliff.
    c. The cliff from single-celled eukaryotes to multicelled eukaryotes is the third.

2. Artificial prebiotic experiments were designed to mimic what might have occurred before life appeared.
   a. Stanley Miller's seminal prebiotic experiment produced several amino acids found in living organisms.
   b. Many other prebiotic experiments followed Miller's, and many other chemicals found in living cells have been produced by these prebiotic methods.
   c. These experiments were based upon the assumption that the Earth's early atmosphere, practically devoid of oxygen, was reducing. Considerable doubt now exists if this was, in fact, the case.
   d. No life has been produced by any of these experiments.
3. Roadblocks to the spontaneous origin of life
   a. The interdependence of proteins on DNA patterns and the replication of DNA that is dependent on proteins.
   b. Patterns for RNA also are stored in DNA, and RNA requires RNA polymerase (another enzymatic protein) for its manufacture.
   c. Did RNA come first with proteins and DNA evolving later? If this happened, RNA would have to act as its own replicating catalyst and also be able to act as its own repository of genetic information. No RNA with these characteristics has been discovered.
   d. Even though all four bases in RNA (guanine, cytosine, adenine, and uracil) have been recovered in the prebiotic experiments, no way has been found to produce ribose in any significant quantities. Without ribose, there can be no RNA.
   e. Using these prebiotic experiments, no nucleotides containing adenine or guanine have been produced. Without them, there can be no RNA.
   f. When attempts to synthesize RNA in the presence of protein enzymes are made (which is cheating from a prebiotic point of view), the reactions needed to produce RNA are inhibited when a 50-50 mixture of right-handed and left-handed nucleotides is used. But in prebiotic conditions, if nucleotides could have been produced, right-handed and left-handed nucleotides would have been produced in equal numbers.
   g. Even if all RNA obstacles are removed, it still could not form biological proteins because all the amino acid building blocks (except glycine) produced under prebiotic conditions would be a 50-50 mixture of right- and left-handed configurations, and RNA would need ribosomes to manufacture the proteins.
   h. Homochirility is one of the biggest problems for origin of life chemists, because there is no random way for exclusively right-handed nucleotides to form or exclusively for left-handed amino acids to form both under prebiotic conditions.
   i. All of the prebiotic experiments were conducted in confining physical systems by smart men, not open systems like the Earth's early atmosphere and oceans. Therefore, any chemicals of life forming in the open seas would have been separated by dilution.

4. RNA is considered by many scientists to have appeared first. The average gene of E. coli is about 999 nucleotides long. Randomly, there would be eight different kinds of RNA nucleotides in primordial soup, four right-handed and four left-handed. To produce one gene 999 nucleotides long would require 8^{999} trials, assuming that no combination of sequences was tried more than once. This number in the base 10 is equal to 1.537×10^{902}. In other words, only one mRNA molecule representing only one gene 999 nucleotides long would be found in 1.537×10^{902} trials.
5. The time constraints of even billions of years is far too small for random formation of even one gene, much less about 4,300 more in primordial soup. Remember however, that life appeared (geologically speaking) soon after the Earth cooled enough so that life could exist. That happened about 700 million years after the Earth cooled significantly and not billions and billions of years later. The first evidence of life is estimated to be about 3.5 to 3.8 billion years ago.
6. The second law of thermodynamics will allow order to be produced from disorder only when the forces released have the proper intensity and direction. Because the forces of evolutionary theory are uncontrolled for intensity and direction as far as the production of the first life is concerned, order cannot evolve from disorder. Without the very first life form, biological evolution's progress from there becomes impossible.

Chapter 10

The Theory of Biological Evolution

Biological evolution, as proposed by Darwin, assumes that at least one living organism exists at the outset. It is an attempt to explain the origin of biodiversity after simple life forms appeared and how biota became more complex. This formed the basis of Darwin's book *The Origin of Species*.

Now let's go back to Darwin's story. After returning to England, he married his cousin, Emma Wedgwood.[1] Prior and continuous returns on investments of stocks given to Charles by his father and a trust fund set up for Emma from her father made the young couple financially independent at the time of their wedding. This annual income of about £1,000 enabled Charles to study the hundreds of specimens he had collected on his round-the-world voyage and to contemplate their meaning.[2] Being financially independent also allowed him time to father ten children, six boys and four girls.[3] Emma Wedgwood was Uncle Jos's daughter, the same Uncle Jos who had helped Darwin to convince his father not only to allow him to go on the voyage, but also finance it.[4]

Soon after his return to England in 1836, Darwin's journal, *The Voyage of the Beagle*, describing his trip was published and immediately became a bestseller.[5] Soon thereafter, he began studying his many biological specimens. He continued this for about ten years. When he thought he was almost through, he took down one of the last bottles that contained barnacles preserved in spirits (alcohol). He became so engrossed in the study of barnacles that he studied these specimens and other barnacles that he obtained later for another eight years.[6] We will learn the reason for his prolonged study in a later chapter.

Remembering the finches on Galapagos Islands and how they varied from those on the mainland of the South American continent, Darwin began to think that the finches on The Galapagos had descended modified from those on the mainland. This was based on the observation that the Galapagos Islands are of volcanic origin and therefore were younger than the mainland 600 miles to the east[7]. Therefore, the few birds that originally colonized these islands had originated on the mainland and had arrived on the Galapagos isles probably by accident, having been blown off course by a storm.[8] These birds, Darwin thought, must have descended modified from the few original colonizers. Then, the notion came to him that maybe all biota on the Earth (plant or animal) had descended modified from a few original individual organisms, or maybe even one. For a time he believed this to be true but could not find a cause to drive this idea along until he happened to read Thomas Malthus' essay, *The Principle of Population.* The essay pointed out the problems that would exist if the human populations were to double rapidly. The human populations are held in check by the resources of food and space available, and by other problems such as premature death and other limiting factors. Think, what would happen if every dandelion seed, every fish egg, or frog egg resulted in a mature organism ready to reproduce again? It was the reading of the essay by Malthus that spawned Darwin's thought for descent with modification via natural selection.[9] It was the competition between all living forms for food and space that provided the impetus for the changes. Any individual organism that possessed even a slight advantage to compete for food and space better than another would have a better chance not only to survive, but also leave more descendants with the same advantage. This advantage of supremacy in survival, even if only slight, would give it preference in a world of competition over a period of time, where selective pressures remained constant. Other individual organisms not containing the beneficial trait would tend to die out and thereby become extinct.[10]

Darwin had seen what man could do by controlling mating of animals and pollinating of plants. In fact, he had experimented with pigeons and noted the changes that he could produce between one generation and the next by controlling which birds were allowed to mate.[11] From these changes that he could produce, Darwin believed that, over time, competition would automatically choose which organisms would reproduce. This was because in all selective pressures imaginable, the fittest would gain supremacy over the less fit. He thought that this would cause beneficial traits to accumulate in various individual lineages of organisms so that they would gradually change into a different species or divide into two. This idea would explain the changes that he thought had occurred with the finches on the Galapagos. He called this process operating in the wild by the term natural selection; others have called it survival of the fittest. Because the changes between one generation and the next are so slight, he realized that if natural selection was involved in the preservation of these very minute changes, millions and millions of generations would be required for beneficial traits to accumulate to produce the huge diversity of biota living on Earth. It was Charles Lyell's great work, *Principles of Geology* volumes I and II, which Darwin had read on board ship that had given him the idea of very long spans of time needed for the many millions of generations to occur. Lyell's third volume was waiting for Darwin on his return to England.[12] Darwin therefore concluded:

1. All living organisms produce more offspring than can possibly survive due to the limits of supplies of food and space.
2. Between various organisms of the same species, there are individual variations.
3. If one of these variations happens to bequeath on an individual a slight advantage to survive, such as color, speed, sight, flight, intelligence, etc. that individual will tend to leave more progeny like itself.
4. Darwin postulated that over millions of generations these small advantages would tend to accumulate, resulting in big changes. These big changes would ultimately lead to the evolution of new species; through the process of natural selection.

For several years, Darwin struggled with the notion of natural selection producing new biota over eons of time. He shared his ideas with a few of his close scientific friends. Toward the end of his nearly 23 years of study, it became clear that Alfred Russell Wallace was proposing a similar theory. Not to be preempted by Wallace, Darwin's friends urged him to quickly write his now-famous book *The Origin of Species*. It was an immediate success. The first printing sold out the day it came off the press. Though his theory has produced much controversy throughout the many decades since that first publication, it has become well-impregnated in modern science. Many scientists consider the theory of evolution to be beyond the theory stage (to what one would almost call a law, like Newton's three laws of motion). However, nothing should be taken for granted. Truth always bears inspection.

A question then follows: do new species arise in nature today by the same forces? In the last paragraph of *The Origin*, Darwin implied that they did. The concept that new species have arisen in the past and that more new species are arising now has come to be known in some circles as Darwin's special theory. When the concept of the origin of life (chemical evolution) is added to the concept of biological evolution, this is sometimes referred to as Darwin's general theory, yet Darwin did not consider chemical evolution in *The Origin*.

Many plants and animals appear closely related because they have similar characteristics. It doesn't take an experienced botanist to notice that citrus trees, be they orange, lemon, or grapefruit, look very much alike. In fact, scions, or buds, taken from all three different citrus fruit trees can be grafted onto the same tree so that it will bear all three kinds of fruit. It is also easy to see the similarities between dogs, coyotes, foxes, and wolves. Darwin saw the similarities that exist in many living organisms. As mentioned previously, he remembered the similarities and differences between the finches that lived on the continent of South America and those that lived on the Galapagos Islands, and having concluded that those finches on *The* Galapagos had descended from those on the mainland, he came up with the idea that all groups of similar organisms were related to each other and had descended modified from a common ancestor. He believed that living species changed over time, very long periods of time, by dividing themselves into two or more species. Hence, the name he chose for his book, The *Origin of Species*.

The significance of chromosomes as the basis of inheritance was recognized independently by Walter S. Sutton and Theodore Boveri in 1902.[13] Thomas H. Morgan discovered how genes are transmitted through the action of chromosomes.[14] It was McClintock and Creighton who discovered the exchange of genetic information during meiosis (See Chapter 6). Hugo de Vries and others around 1905 began to understand the significance of mutations as another cause of variation from one generation to the next.[15]

From this point of view, we will start to examine evolution right where Darwin started, that is, with sexually reproducing multi-celled eukaryotes. He knew practically nothing about prokaryotes and single-celled eukaryotes, as very little was known about them in his time. It must be emphasized that Darwin built his whole theory on his observations of multi-celled sexually reproducing eukaryotes. Now we understand the causes for change between one generation and the next, namely the triad of meiosis, sexual reproduction, and mutations, which he didn't, because their discovery came after his time. But remember, according to evolutionary theory, changes must accumulate over long periods of time to produce enough genetic modification to cause, not only new species to evolve but also new kingdoms, phyla, classes, orders, families, and genus to arise. If true, all biota descended modified from one or several primordial organisms.

Darwin noted that there were slight differences in the same species between one generation of sexually reproducing organisms and the next. Actually, he gave many examples to show that when humans control the breeding of domesticated plants and animals, man can develop improvements in those plants or animals in multiple successive generations. Later, Darwin called these changes caused by man's control of the breeding of plants and animals **variation under domestication**. The changes that he noted taking place in the wild between one generation and the next, he called **variation under nature**.

In his travels, he noted that all of life seemed to be in a "Struggle for Existence." This is the title of the third chapter in The *Origin of Species.* We see this all around us. In fact, we don't pay much attention to this phenomenon as we go about our daily lives. Spiders catch insects in their webs to eat, birds eat spiders, snakes eat birds, coyotes eat snakes, etc. In the water some large insect larvae eat small fish and little fish eat mosquito larvae and big fish eat little fish. The examples are endless.

Darwin, in this chapter, proposed that this struggle directed descent with modification in nature. Two sentences from this chapter seem to summarize what Darwin thought: *"Owing to this struggle for life, any variation, however slight, and from whatever cause proceeding, if it be in any degree profitable to an individual of any species, in its infinitely complex relations to other organic beings and to external nature, will tend to the preservation of that individual, and will generally be inherited by its offspring. ... I have called this principle, by which each slight variation, if useful, is preserved, by the term of Natural Selection, in order to mark its relation to man's power of selection."*

Man directs descent with modification under domestication by controlling the breeding of plants and animals. In the wild, Darwin thought that the force of natural selection directed the descent with modification. All living organisms are in a struggle for existence. Those that survive and leave the most progeny are successful and have had an advantage over those that don't. The advantage may be ever so slight. Nevertheless, even a small advantage is helpful in survival. According to Darwin's theory, after thousands or even millions of generations, many slight, beneficial traits will accumulate by this selection process, which will gradually perfect the organism. Those with a slight disadvantage will become extinct. Two sentences from Chapter 4, which he titled Natural Selection, exemplify his thinking: *"It may metaphorically be said that natural selection is daily and hourly scrutinizing throughout the world, every variation, even the slightest; rejecting that which is bad, preserving and adding up all that is good; silently and insensibly working, whenever and wherever opportunity offers, at the improvement of each organic being in relation to its organic and inorganic conditions of life.*

We see nothing of these slow changes in progress, until the hand of time has marked the long lapses of ages, and so imperfect is our view into the long past geological ages, that we only see that the forms of life are now different from what they formerly were."

Darwin's statement refers to natural selection's function as a metaphor. He saw natural selection as an omnipresent metaphoric judge "*daily and hourly scrutinizing, throughout the world, every variation, even the slightest; rejecting that which is bad, preserving and adding up all that is good.*" From this, we can see from Darwin's point of view that natural selection is a process that makes non-random choices based on competition between two or more living organisms. Organisms, or groups of organisms, possessing characteristics of superior competition for food and space would be able, Darwin thought, to prosper, while forcing into extinction others that possessed inferior characteristics needed for survival. This process obviously is blind, has no memory, or goal, and acts without any intelligent guidance. What is not so obvious in Darwin's seminal statement about the force driving his theory is that natural selection, or survival of the fittest, can act only as a judge to make choices between living organisms that have differences related to survival. It can never make a selection between organisms with equivalent abilities to survive, nor can it make a selection between silent mutations in living organisms because they do not produce any differences.

When Darwin proposed his theory in 1859, the science of genetics had not been developed. Gregor Mendel's written records of his work on genetics in the 1860s were not discovered until 1900.[16] Nevertheless, Darwin could see that there were slight variations in succeeding generations that he called descent with modification. He postulated that all living things were in a constant state of flux, and species were not immutable, but were constantly undergoing slight change. By extrapolating from the few changes that he could visualize occurring in short periods of time, he came to believe that characteristics in a specific species now might have been completely different a million generations back. Gradually, over millions of generations, these changes could accumulate to the point that, say, fish could become progenitors of amphibians. By accident, the simple become more complex as millions of generations pass by.

It is now common knowledge that variations produced by the triple-mixing process of meiosis and sexual reproduction alone could not be responsible for fish to become the progenitors of amphibians, even with millions of generations in between, because pure meiosis and sexual reproduction only mixes genetic material without adding any. After Darwin's time, with the discovery of mutations and the crossing-over element of meiosis, scientists thought that they could see how new genetic material could be added or old genetic material subtracted or altered in the genes of a given organism by mutations. The accumulation of beneficial mutations, in their opinions, theoretically could allow one class, such as fish, to become the progenitors of amphibians, such as frogs. While the triple-mixing process of meiosis, sexual reproduction, and mutations are completely random events, natural selection is not. After living organisms came on the scene, random mutations are chosen by it to survive or die. Richard Dawkin's book, *The Blind Watchmaker,* points out that natural selection, or as he refers to it, "Cumulative Selection," is not random. Natural selection does not operate by chance because it chooses beneficial mutations to live and deleterious mutations to die.[17] With the addition of mutations into the evolutionary theory, it is now referred to as the Neo-Darwinian synthesis.[18] When this synthesis was first proposed, the genetic code, composed of 64 codons and genomes with millions of base pairs, was not known. Consequently, the number of random mutations thought needed for evolutionary progress to occur was much less than is known to be required now. The Neo-Darwinian synthesis depends upon thousands, and in some cases, millions of random **beneficial mutations** to change the DNA at just the right place and in perfect sequence for natural selection to preserve in order to evolve one class of life into another, as with a fish evolving into a frog. As you will learn in succeeding chapters, the majority of mutations are deleterious, damaging the individual organism in which they have occurred. Some mutations are silent causing no change. Natural selection cannot detect a silent mutation, because a silent mutation does not change the protein that the mutated codon specifies. It is imperative that a change in function is required to trigger the choosing process of natural selection to preserve or reject a given organism or biological process. When two or more biological organisms or processes are in equilibrium with each other as far as competition between them is concerned, natural selection cannot choose between them. However, if one of the competitors has a change in function from any cause, which has disturbed the equilibrium then and only then can natural selection, the non-random judge of evolutionary theory, make a choice to preserve or reject the change.

Beneficial mutations, if they occur at all, are extremely rare. All mutations are random events as to both placement in the DNA and as to timing when they happen. Evolutionary theorists believe that the postulated beneficial mutations would be retained by natural selection to flourish, even in a changing environment.

Here we go with the finches on the Galapagos Islands again! Darwin noted that the finches on the islands differed from those on the mainland. Over thousands of years, through the process of natural selection, he thought they had accumulated beneficial characteristics that helped them adapt to their new environments. This gave them characteristics different from their progenitors on the mainland. For Darwin, these changes exemplified the phenomenon of *variation under nature and descent with modification*, preserved by natural selection. Because no two individuals of the same species have exactly the same DNA sequences, except for identical twins, there is a lot of latitude for the triple mixing processes of meiosis and sexual reproduction to bring out various different characteristics hidden in the DNA. Although Darwin did not know about these triple-mixing processes, he did see that there were slight differences between generations even of the same species. Some of these characteristics he thought would be beneficial and natural selection would choose them to survive; in others it would discard. Darwin postulated that beneficial characteristics could accumulate over time and aid a group of organisms in adapting to a new environment. As far as the finches on Galapagos are concerned, the birds stayed birds!

We must pause here to reflect and enumerate the basic implications of evolutionary theory. We have started where Darwin started by analyzing how changes occur between succeeding generations of biota. This, of necessity, means that we have begun where he did with multi-celled, sexually reproducing eukaryotes. When Darwin wrote *The Origin*, very little was known about prokaryotes and certainly nothing was known about their single, circular chromosomes. Also nothing was known about the paired chromosomes and organelles located in both single-celled and multiple-celled eukaryotes. Darwin based his theory almost exclusively on his study of sexually reproducing eukaryotes. We now know that this last group triple-mix their genetic material with each succeeding generation. He knew nothing of the millions of base pairs required for even the simplest life forms. He knew practically nothing of genetics, but he could see subtle differences between the generations of the same species. From what he could see, he extrapolated backward in an effort to try to understand what he could not see. We, from our vantage point of much greater knowledge, must metaphorically follow in his cognitive footsteps to understand better what is involved in his evolutionary theory. From the multi-celled, sexually reproducing eukaryotes, we must work our way down to the single-celled eukaryotes, then down to the prokaryotes (the top down approach), reconstructing in our minds how life started simple and became complex. Darwin theorized that living biota started simple and gradually became more complex as beneficial traits slowly accumulated over millions of successive generations. From this we can see that his theory includes three basic ideas, "species are not immutable," it provides an explanation for biodiversity, and natural selection acts as the guiding force.

But how is a species defined? In the first paragraph of Chapter 2 of *The Origin of Species,* Darwin said: --- *"Nor shall I here discuss the various definitions which have been given of the term 'species.' No one definition has yet satisfied all naturalists; yet every naturalist knows vaguely what he means when he speaks of a species."*

Scientists still have difficulty defining what a species is. The high school biology text book, *Biology, Visualizing Life*, by Johnson has the simplest definition. It states, "*A species is a group of organisms that look similar and can produce fertile offspring in their environment*."[19] Scientists who classify living things are taxonomists, and the classification of living things is taxonomy. The biology text's definition uses the words, "looks similar." Similar looks is a simple way of comparing morphology. Morphology describes shape and structure of an organism. Another method used to define species implies that two separate species will not interbreed a type of reproductive isolation. However, it is possible to cross domesticated cattle with the North American buffalo. This produces what is known as cattalo. A pair of cattalos also can produce fertile offspring, which would come under the heading of a new species, as the biology textbook defines them. This was reported in Wonders of Animal Life edited by J A Hammerton (1930).

But it would be a considerable stretch of our imaginations not to think that a cow and a buffalo are not two separate species, simply by contrasting their respective morphologies. The same observation has occurred among at least two of the supposedly 13 separate species of Darwin's finches on the Galapagos Islands. These supposedly different species have been noted to interbreed and produce vigorous offspring, which also have been found to reproduce. Even here, there is not reproductive isolation as was formerly believed. If they had evolved apart, the genetic distance between the two was not as much as had been previously postulated. [20] Then there are the fossils. Because they are dead, there is no way to apply to them directly the test of producing fertile offspring. But obviously, they did or they would not have existed. However, they present themselves in less than optimal condition. Therefore, consideration can be given mainly to the morphologies of their remains.

Just as Darwin pointed out in the quote from *The Origin* mentioned above, there was no perfect definition of a species then nor is there one now which satisfies every biologist. However, it is possible to conclude that all of these categories of taxonomy suggest a hierarchy of typology in which there is no overlapping of the classes as noted between, say, fish and amphibians. Nature's divisions do not seem blurred and indistinct, except possibly at the genus or species level. Taxonomists have divided all biota into kingdom, phylum, class, order, family, genus, and species. However, living protoplasm always has, and always will, reside in species. Darwin, in *The Origin*, stated that *"every naturalist knows vaguely what he means when he speaks of a species."* The ultimate method of separating two similar living species will come from the analysis of the DNA in their respective genomes. The foreseeable future very likely will produce this type of evidence. So here we are well over a century after Darwin, trying to understand how species appeared and still species defies a complete scientific definition for all cases. Even though a perfect definition of a species does not exist, for practical purposes every species can be classified as belonging to a kingdom, phylum, class, order, family and genus, yet every organism resides only in a species.

SUMMARY

1. All living organisms produce more offspring than can possibly survive due to the limits of supplies of food, space, and presence of predators.
2. Between various organisms of the same species, there are individual variations.
3. If one of these variations happens to bequeath an individual with a slight advantage to survive, such as color, speed, sight, flight, intelligence, etc., that individual will tend to leave more progeny like itself, as long as environmental conditions favoring that trait remain unchanged.
4. Darwin thought that over millions of generations these small advantages would tend to accumulate, resulting in big changes. These big changes would then lead ultimately to the evolution of new species, he believed, through the process of natural selection.
5. The question in Darwin's mind was: what could bring about change of species in the wild?
6. He came up with the notion that he called natural selection. Some call it survival of the fittest.
7. He reasoned, if any beneficial variation took place in an organism in the wild, it would help that organism to compete more effectively than competitors, for food and living space and thereby leave more offspring with the same beneficial variation.
8. In any organism a genetic variation that was deleterious would make it more difficult for that organism to compete in its environment, thereby leaving fewer, if any, offspring.

Eventually, organisms in this latter group would end in extinction because of their inability to effectively compete for food and living space.

9. Darwin based his theory on the slight changes that he noted between succeeding generations of sexually reproducing multi-celled eukaryotes of the same species. He knew practically nothing about prokaryotes when he wrote *The Origin*, because virtually nothing was known about them at that time.
10. Even though meiosis followed by fertilization, is a powerful mixer of genetic material that can cause considerable modification between succeeding generations of sexually reproducing multi-celled eukaryotes of the same species, they could not bring about the changes needed to change a fish into a frog, even over millions of generations. This is because they only mix existing genetic material but add no new genetic material.
11. With the discovery of mutations, scientists then believed that mutations provided a mechanism that would produce the changes needed. They thought that random mutations selected or rejected by natural selection would allow beneficial traits to accumulate by the addition of new genetic material to the genomes of mutated organisms. After millions of generations, mutations filtered by natural selection would produce the changes necessary to evolve one class of organism into another. Theoretically many scientists believe that these changed genomes can be traced back to some primordial ancestors. With the addition of mutations into evolutionary theory, this combination is now known as the Neo-Darwinian synthesis.
12. To produce the thousands of differences in the number and sequence between the codons of any two taxonomic classes (which theoretically can be traced back to some common primordial ancestors), there must of necessity be thousands, or in many cases, millions of mutations to construct two different lineages. In order for survival of the fittest or natural selection to preferentially select a given mutation, it must be beneficial.
13. To change one organism into another over many thousands or millions of generations, a series of beneficial mutations, though random both as to placement and timing, must happen by pure chance at just the right place in the DNA with perfect timing and sequence. Silent or neutral mutations though random as to placement and timing, may become beneficial or deleterious in later generations depending upon mutations that may occur later.
14. All fatal, deleterious mutations would be eliminated by natural selection in extention.
15. The evolutionary theory as proposed by Darwin in 1859 in his book *The Origin of Species* is believed by many scientists to have moved a little beyond theory to almost a law. In their minds, it is certain.
16. It is only by a random mutation that a change in the DNA can be presented to natural selection to determine if the mutation is deleterious, silent, neutral, or beneficial to the organism involved. This phenomenon can occur only after the appearance of the first life form, most likely a prokaryote as there is no such thing as a mutation in non-living material.
17. To date, there is no universally accepted definition of species that applies in all cases.

CHAPTER 11

Confining Boundaries

From the Big Bang to the present, the age of the universe according to the WMAP Satellite measurements, was 13.772 + or – 0.059 billion years. This was followed by the European Space Agency Satellite measurements of 13.798 + or- 0.037 billion years. The length of the six Creation Days in cosmic time was calculated by using the stretched CMB waves as figured in chapter 8. When cosmic time was converted to Earth time, Creation week took 13.705 billion years to complete. However, initially after this giant explosion, life as we know it could not have survived in this hot environment. It is now estimated that our galaxy, our solar system, and our planet did not form until billions of years after the Big Bang. It is common knowledge that the inner layers of the Earth are still molten, as exemplified by the lava that flows from volcanoes. This means the most external layer, or shell, of Earth had to cool enough before liquid water could condense and life could survive. It is believed that these parameters were reached approximately 4.5 billion years ago.[1] Evidence of simple life forms have been found in Earth's oldest rocks, which indicates that life appeared about 3.5 to 3.8 billion years ago.[2] This means life had to arise, if by natural means, in a relatively short time, geologically and evolutionarily speaking. Keep these two questions in mind as you read this chapter. First, is there enough time; and second, is time a confining boundary? These two questions will be explained in more detail in chapter 12.

When analyzed carefully, Darwin's whole theory rests entirely on circumstantial evidence. This is because no one has ever witnessed a new life arising or visualized an established life divide into two separate species. Science always must consider circumstantial evidence in two forms: positive and negative. Positive circumstantial evidence can show that something might have occurred, but can never show that it did. However, negative circumstantial evidence is more powerful. It can show that something has not occurred or could not occur at all. As a result of this logic, one well-documented bit of negative circumstantial evidence must always take precedence over a large amount of positive circumstantial evidence. For the most part, Darwin built his whole theory on positive circumstantial evidence.

A quotation from Stephen Hawking, describes negative circumstantial evidence to a T. He recognizes this logic in his short discourse about scientific theories in his book, A *Brief History of Time*. He calls a theory a model existing only in our minds. On pages 9 and 10, Hawking says, *"A theory is a good theory if it satisfies two requirements: It must accurately describe a large class of observations on the basis of a model that contains only a few arbitrary elements, and it must make definite predictions about the results of future observations.... Any physical theory is always provisional, in the sense that it is only a hypothesis: you can never prove it. No matter how many times the results of experiments agree with some theory, you can never be sure that the next time the result will not contradict the theory. On the other hand, you can disprove a theory by finding even a single observation that disagrees with the predictions of the theory."* (Negative circumstantial evidence.)

One thing that might lead us to believe that all living things are related and have descended modified from one original organism is the genetic code in the DNA. As far as we now know, every living organism from bacteria to man and all living forms in-between, in all five kingdoms, use the same basic genetic code. So, does that show descent with modification from a common ancestor? It could, but doesn't necessarily have to do so. However, it is a powerful bit of positive circumstantial evidence in its favor.

The alphabet is a code, but we have gotten so used to it that we don't think of it as a code. We use it as a written code for verbal sounds that we speak. Just because our alphabet is a code that an English author can use to write a book or a Spanish author can use to write a poem, does not mean that there cannot be separation between the two rather than an absolute continuum. The subject matter of the English author could be about how to analyze the air foil on the wing of an airplane, and the Spanish author might be poetically telling his readers about barnacles. Just because the alphabet, which is a code for sounds, can be used for a myriad of subjects in different languages, doesn't prove that all subjects are related. The same also can be true of the genetic code or alphabet of life.

As Darwin traveled around the globe, he observed many different forms of biota in many different environments. From these observations and the many specimens that he collected on his trip, he tried to arrive at some kind of overall explanation for what he had observed. Like Mendeleev, who devised the periodic table for chemicals, Darwin tried to categorize all of biota in an all-inclusive theory. When all of the evidence for his conclusions regarding the complexity of biota are metaphorically laid out, it is difficult to reject his explanations. It is difficult to reject the idea that much of biota is not related, when in fact it is possible at least in the plant world, to graft different species of trees onto the same tree. For instance, oranges, lemons, tangerines, and grapefruit can all be grafted onto the same tree and thrive. Each of these species must be very closely related. In the animal world foxes, coyotes, wolves and dogs look similar. Many other examples could be cited.

Darwin built his entire theory on positive circumstantial evidence. As just described, circumstantial evidence comes in two forms: positive and negative. Negative circumstantial evidence is always more powerful than positive circumstantial evidence, as it only takes one bit of negative circumstantial evidence to contradict many bits of positive circumstantial evidence. An illustration from a crime that took place in 1988 will be cited to show how powerful negative circumstantial evidence can be. The following true story was gathered on line from several news sources—DallasNews.com, salon.com, and Wisconsin Innocence Project.

One day Christopher Ochoa and Richard Danziger (both roommates and both employees of a Pizza Hut in Austin, Texas), happened to stop by another Pizza Hut in that same city. Several weeks before their visit to this store, the manager of this Pizza Hut had been raped and murdered as she was opening early one morning. Before the two men happened by, there had been no suspects and very few if any leads. Employees at this second store where the crime had been committed thought that Chris and Richard looked suspicious. They called the cops! Later Chris was picked up by the Austin police at his workplace. After two days of prolonged questioning by more than one tenacious detective, Chris finally admitted his part in the crime and also implicated his roommate as well. This was in 1988. Both were sent to prison for life.

In prison, Chris took advantage of the educational opportunities offered, which ultimately led to his completion of two years of college. During his course of studies, he came across the Wisconsin Innocence Project. In 1997 Mr. Ochoa contacted them. Students at the University of Wisconsin Madison, School of Law participate in this project and try to discover any unused or ignored evidence that might exonerate someone from prison. The law students found that there was DNA evidence still available from the crime scene that might prove Chris and Richard innocent. The DNA technology had advanced markedly in the decade from 1988 to 1997. When this DNA evidence using the advanced technology was applied to this case, it proved that both Chris and Richard were in fact innocent. Both were released from prison.

In retrospect one wonders why Chris signed papers incriminating himself for a crime that he did not commit? At the time of his confession, Ochoa was a young 22-year-old Mexican-American Pizza Hut worker without much if any legal savvy. He was grilled for two full days by detectives who used extreme measures to get Chris to confess. One told him that he knew Chris had done it. He lied when he said that Richard was in the next room, and that he was about ready to rat on Chris. One or both of the detectives told Chris that "white guys always walk" and that Hispanics always get the needle. One of the detectives pounded the table, pushed on the place on Chris's arm where the needle would be inserted, and at one point threw a chair at Chris barely missing his head. He was told that if he went to trial and was found guilty, he would certainly be placed on death row. They showed Chris pictures of death row. However, he was told if he confessed to the crime, he would get life in prison. Worn to a frazzle both physically and psychologically and thinking that life in prison would be better than death, Christopher Ochoa confessed to a crime that he didn't do and also implicated Richard, his non-hispanic roommate and fellow worker.

Even before the DNA tests proved Chris and Richard innocent, other evidence had come to light. In 1996 a felon by the name of Achim Josef Marino, serving time in a different prison for a another crime, wrote a letter to the Austin chief of police and Travis County District Attorney Ronnie Earle, confessing to the Austin Pizza Hut crime. Nothing happened! So two years later Mr. Marino sent another letter to then former Governor George W. Bush. Again no response! Marino even told in his letters where police could find two Pisza Hut money bags and keys to the store where the crime had taken place. Later, this evidence proved true. After the DNA test had exonerated Chris and Richard, it also implicated Marino.

In January 2001, after 12 years in prison for a rape/murder he did not commit Christopher Ochoa was released from prison. Chris sued the Austin police for poor handling of the case and received a financial settlement. He used the money to enter college and received a degree in Business Administration. In 2003, Chris entered law school at the University of Wisconsin, Madison. In 2004, he became a student in the Wisconsin Innocence Project, the same program that had helped to prove his innocence. In May of 2006, he graduated with a law degree from that same University.

One very sad note to this saga concerns Richard Danziger. While in prison, another inmate mis-took Richard for someone else. This felon beat him severely, kicked him in the head repeatedly, which caused severe brain damage that required surgery. He will never be the same!

This is a classic example of how positive circumstantial evidence, which even included a signed confession, was proved wrong by one important bit of negative circumstantial evidence. The tenacious detectives remind us of how determined many evolutionary scientists are to prove Darwin's theory true. Metaphorically, they stomp and scream and throw imaginary chairs at anyone who disagrees with their cherished paradigm and all the while they ignore blatant negative circumstantial evidence, some of which is confirmed by more advanced technology. They even go to court with this evidence which is presented in such a way as to make any other approach to this question appear to be held only by the ignorant. This is reminiscent of the educated elite of Galileo's day who refused to look through his little telescope to see what he saw. This goes on in spite of multiple examples of powerful negative circumstantial evidence that has surfaced in the past few decades as technology has advanced. No matter how many pieces of positive circumstantial evidence that can be garnered for a given scientific theory, one significant piece of negative circumstantial evidence will show that the theory is either false or needs to be drastically overhauled. As we proceed to examine Darwin's theory of evolution in the remaining chapters, please keep in mind these two forms of circumstantial evidence. And remember truth always bares inspection!

Stephen Hawking, quoted earlier in this chapter, said *"A theory is a good theory if it satisfies two requirements: It must accurately describe a large class of observations on the basis of a model that contains only a few arbitrary elements, and it must make definite predictions about the results of future observations... He also said, "you can disprove a theory by finding even a single observation that disagrees with the predictions of the theory."* From this we can see that there may be many bits of positive circumstantial evidence in favor of a theory, but negative circumstantial evidence against it must be missing or else the theory must be discarded or drastically overhauled. This notion applies to biological evolution, as Darwin visualized it, as well as to Neo-Darwinism, as it is understood by its proponents today.

The few arbitrary elements composing a model of the theory of evolution can be described in three sentences. Life started simple and became progressively more complex as beneficial characteristics slowly accumulated over eons of time in millions of succeeding generations of organisms driven by natural selection's ability to choose the fittest. As complexity accumulated, many divisions occurred, as various lineages became separated and diversified. This produced myriads of new species. It is very important to remember that evolutionary theory claims to be self-driven and therefore can have no memory, goal, or intelligent guidance. According to Hawking, the above brief summation of Darwin's theory describes a large class of observations with a few arbitrary elements. In one fell swoop, it attempts to explain how every living organism, including us humans arrived here. But Hawking also says that a good theory also must make predictions about the result of future observations and must not have any evidence against it. In other words, negative circumstantial evidence must be absent.

The scientific method has been centuries in the making. Though it has a general form, it seldom is followed precisely. First, through observations of natural phenomena or experiments, data are collected. Next, attempts are made to discover if patterns are present. If a pattern is discovered in the data, a hypothesis or theory is proposed. From this, predictions are made as to what will be found in future observations or experiments. The hallmark of any successful scientific theory is that predictions are forth-coming from it, which when tested by observations, or experiments, are proven true. These predictions frequently lead to new discoveries. However, if even one prediction is proven wrong (negative circumstantial evidence) the hypothesis or theory must be abandoned or drastically rearranged.

As a classic example, a scientist in the nineteenth century who collected data that had been gathered by his predecessors for centuries before him was Dmitry Mendeleyeev, a Russian chemist. He organized the then known 63 elements into what now has become the periodic table of the elements. He noted the similarities and differences between the various elements that had been identified up to that time. From these observations he discovered patterns from which he compiled his table in such a way that it became obvious that there were many slots missing where elements should be located. These missing elements were begging to be discovered. Subsequent chemists isolated these elements which filled in the blanks and others added elements not even anticipated by his table. Among these were the noble gases discovered by William Ramsey. His discovery added an entire column to the far right of Mendeleyeev's table. For this discovery Ramsey received the Nobel Prize in 1904.

Since 1871, when Mendeleyeev first published his periodic table, scientists have readily accepted it because it not only systemized the data known at that time, but also made specific testable predictions about the missing elements waiting to be isolated.[3] In fact, Mendeleyeev made predictions that other elements would be found to fill in the blank places in his table. He predicted what their chemical properties would be. Within 20 years after his predictions many of these elements were isolated.[4]

Darwin consciously, or unconsciously, followed this method. First he collected the data on his circumnavigational trip. By the time he returned to England he had chronicled 1,383 pages of geology notes and 368 pages of zoology notes. He had cataloged 1,529 species preserved in spirits (alcohol), and labeled 3,907 skins, bones, and miscellaneous specimens. He had also brought a live baby tortoise from The Galapagos Islands. His diary contained 770 pages.[5]

For nearly 20 years after Darwin returned to England, he contemplated the data that he had collected. From this he thought that he could see a reoccurring pattern. From this he proposed his theory of evolution and made predictions. So what predictions did he make and are other predictions implied? Is there any negative circumstantial evidence? The answer to both is yes! The first test is implied in the theory itself. Three more were proposed by Darwin in *The Origin*.

1. Does natural selection actually work in the wild to benefit one group of organisms and work as a detrimental force against others?
2. Have new species arisen in the wild as a result of the force of natural selection? Darwin said, "*By the theory of natural selection all living species have been connected with parent-species of each genus, by differences not greater than we see between the varieties of the same species at the present day; and these parent-species, now generally extinct, have in their turn been similarly connected with more ancient species; and so on backwards, always converging to the common ancestor of each great class. So that the number of intermediate and transitional links, between all living and extinct species, must have been inconceivably great. But assuredly, if my theory be true, such have lived upon this earth."... "Geology assuredly does not reveal any such finely graduated organic chain; and this perhaps is the most obvious and greatest objection which can be urged against my theory. The explanation lies, as I believe, in the extreme imperfection of the geological record."* (*The Origin,* chapter titled: "Imperfection of the Geological Record").
3. "*If it could be demonstrated that any complex organ existed, which could not possibly have been formed by numerous, successive, slight modification, my theory would absolutely break down.*" (Chapter titled Difficulties on Theory).
4. In the chapter XIII in *The Origin*, Darwin takes up the notion that homologous parts in organs of different species suggest that these species are related. *"This is the most interesting department of natural history, and may be its very soul."* He calls this observation the "very soul" of natural history. The topic of homology will be discussed in chapter 13 of this book.

So let's discuss the first test. Is there evidence that natural selection works in the wild, demonstrating survival of one group due to an advantage, and death of another group due to a disadvantage? Is there evidence that new species have developed in the wild due to natural selection? Answers to these two questions should help us understand whether natural selection, over eons of time, has worked as the driving force of the development of new species.

It seems strange that almost one hundred years went by before someone tried to document natural selection or survival of the fittest as it takes place in the wild. A scientist named Kettlewell finally put the notion of survival of the fittest or natural selection to the test. It was well-known that the British peppered moth (*Biston Betularia*) came in two different colors, mottled gray and mottled black. These moths were known to interbreed just like two different colored cats. For the most part, these moths are nocturnal, that is, they fly around at night when insect-eating birds are sleeping. Depending on the color of moth, their camouflage matched the tree bark or rocks on which they roosted during the day. This made them less visible to insect-eating birds that search for food during the day. Part of a century before and after Darwin's publication of *The Origin of Species* in 1859, England experienced the Industrial Revolution. Its main source of energy came from the burning of coal. Over the years, soot from the burning coal in the cities had colored the tree bark and rocks black. As time passed, the mottled black peppered moth matched the black surroundings in the city better than the mottled gray peppered moth and therefore formed the predominant portion of the moth population. Their color camouflage protected them better from hungry birds looking for food in the city. However, in the countryside, where there was no significant amount of deposited soot, the mottled gray formed the majority of the peppered-moth population. Kettlewell trapped hundreds of each colored moth. He placed a small dot of fast-drying cellulose on the under surface of each moth to identify it later as one of his trapped moths. On a given day, he released many of the two different colored moths (that is the light and dark) in the city as well as the countryside. Some days later, he set traps in both places to discover whether these color changes would have any effect on survival rates from the eyes of hungry birds. In the country, the speckled gray moths formed the greatest number of the recaptured moths, but in the city the black moths were in the majority. Before the Industrial Revolution, both colored moths existed equally in city and country. The inference from this experiment was that the color proved to be an advantage depending upon which environment the two colored moths lived. In the cities, black moths had better chances of survival and were able to leave more progeny, and the opposite was true in the country. Natural selection was at work in the struggle for life, preserving organisms with the color advantage.[6] Even though this experiment supposedly proved that natural selection works in the wild, it did not prove that a new species arose by this method in spite of the fact that this selection process went on for more than two centuries. Though two centuries is a relatively short time, evolutionarily speaking, it did not produce a new species of moth, even with these optimum selective pressures. These effects of color change of the moths in the two different environments are easily explained by rules of genetics and meiosis preserved, or rejected, by natural selection. So we can see that it takes longer than two centuries to produce a new species in the wild. In two centuries, man, by carefully controlling the genetic flow, can probably produce a new species or sub-species of dog or cat, but in the wild no rapid changes occur. There seems to be confining boundaries which are difficult for the selecting process of natural selection to cross.

Thus, in just a few generations, the genetic material specifying the jet black (melanic) color of peppered moths increased to predominance among peppered moths living in an environment blackened by soot. During the day, these moths roost on rocks and tree bark. It is obvious that black-colored moths roosting on black-colored objects will be harder to see than light-colored moths by birds that feed on these insects. As time goes by, the percentage of black moths will increase in the black environment as more and more of the light-colored moths are eaten by hungry birds who can spy them easier. As the percentage of light-colored moths decrease, black moths will tend to mate with black moths, producing more black moths. The exact opposite will be true for moths living in light-colored environments. Regardless in which environment these moths lived, the change in survival rates between the two different-colored moths did not produce a new species of moth. In fact, when soot-blackened environments have been cleaned up, the percentages of peppered black and peppered gray moths soon returned to normal, which followed the genetic rules of variation without addition or subtraction of genetic material.

Although Kettlewell's seminal peppered moth experiment conducted in the early 1950s supposedly demonstrated the existence of natural selection in the wild as the driving force of evolutionary change as proposed by Darwin nearly a century earlier, this whole scenario has in recent years come under critical review. This includes not only criticism of the way the experiment was carried out by Kettlewell, but also attacks on the veracity of the man himself. This problem was briefly described in *Science*, June 25, 2004, Vol. 302, No. 5679, pp. 1894-1895. It is actually astonishing to have conclusions of Kettlewell's peppered moth experiment questioned by both evolutionists and antievolutionists alike after decades of its acceptance as the objective evidence of natural selection at work in the wild. The result of this experiment has been the "poster boy" for natural selection, being showcased in high school and college biology textbooks, as well as in scientific books and journals.

Now let's take up the second question. Do new species develop in the wild and specifically as a result of the forces of natural selection? Besides the evidence for speciation of finches on The Galapagos Archipelago, examples of speciation have been studied on the Hawaiian Archipelago as well. These studies have shown that new species of birds have evolved after the original colonizing birds arrived there many millennia before. But the birds stayed birds, just as they did in The Galapagos. In addition, the tremendous bio-diversity of aquatic animals found in such ancient lakes as Lake Victoria in Africa and Lake Baikal in Russia are highly suggestive of local speciation that developed. (*Science,* January 31, 2003, Vol. 299, No. 5607, pp. 654-655).

In Lake Victoria, it is quite evident that new species of cichlid fish have evolved from previous cichlid fish. However, it is the color vision of female cichlid fish in choosing a mate that drives the speciation and not natural selection based on survival of the fittest. Even though this color change produced a new species, it undoubtedly did not help the cichlid fish compete for food and space among predators or competitors. In no case have any examples been documented, even in the fossil record, where one class of organism evolved into another class, like fish having evolved into amphibians such as frogs. The above examples of speciation make it quite clear that Darwin was partially right, at least as far as speciation described in *The Origin* is concerned. However, development of these new species can be explained by the laws of inheritance resulting from meiosis and sexual reproduction, which can bring out different traits in living forms, the patterns of which already existed hidden in the DNA. These changes can occur in the wild without beneficial mutations, similar to how new species of dogs or cats can be produced by man's control of genetic flow.

Remember the genetic distance between one class of biota and another, or one kingdom and another, can only be traversed by the addition of huge amounts of new genetic material as well as rearrangement of existing genetic material. These necessary additions and changes impose confining boundaries for progress of evolution because, according to evolutionary theory, they are totally dependent on millions of perfectly placed and perfectly timed beneficial mutations.

The evidence that this accumulation has slowly occurred in thousands, or even millions, of successive generations as noted in the fossil record is missing. This will be described more fully in chapter 14 of this book.

There may be a speciation process which falls somewhere between those produced by meiosis and sexual reproduction alone and those produced by meiosis and sexual reproduction plus millions of perfectly placed and perfectly timed mutations. The former is exemplified by the speciation of cichlid fish and the latter would probably best be represented by the theoretically postulated evolution that was supposed to evolve single-celled eukaryotes from prokaryotes. Probably in-between these two is a situation called "Speciation by Distance in a Ring Species." An article by this name appeared in the January 21, 2005 issue of *Science* on pages 414 to 415, Vol. 307. This article describes an apparent set of conditions where speciation has occurred in Siberia between two species of greenish warblers (Phylloscopus crochiloides viridanus in western Siberia and Phylloscopus crochiloides plumbedarsus in eastern Siberia).

Imagine a geographic ring with a split located at its northern most position. On one end of the split in the ring is a large population of greenish warblers (Pt viridans). It has been proposed that as thousands of years came and went, this species of bird increased in numbers and expanded into a territory south of the original location in Siberia. From there they later expanded in an easterly direction and finally north again until they came in contact with descendants of the original warblers that had also expanded in an easterly direction from their starting point. The two species are still greenish warblers but the one on the easterly side of the original split in the geographic ring is different from the one on the west. There is no way of knowing on which end of the split in this geographic ring, either west or east, a new species appeared, because the expansion could have been the reverse, that is from east to south to west to north. However, by directly analyzing the amplified fragment length of polymorphism markers, a rather smooth chemical continuum was traced between 105 greenish warblers at 26 different locations proceeding from north to south then to east and lastly north again along the geographic ring.

The analysis of the chemical changes found in these markers located in these birds living in different locations just described on this geographic ring seemed to trace out a smooth chemical continuum between these two species of bird that shows they are chemically related, but will not interbreed. This example is believed to demonstrate speciation forced by geographic separation between these two species of greenish warblers. These chemical changes may or may not have involved some subtle form of mutations. Regardless, the birds stayed birds. Both species were easily recognized as being greenish warblers. Undoubtedly, thousands of years had passed to produce the subtle chemical differences between these two species of birds, yet these two species were confined to being greenish warblers. They had not evolved into mockingbirds or scarlet tanagers, they remained greenish warblers.

The chemical changes witnessed in the greenish warblers, if resulting from mutations, could hardly be classified as beneficial. Even though these two species appear to be related chemically as the geographic ring is traced out, they seem to remain reproductively isolated from each other since they do not interbreed. Regardless, the chemical changes that occurred, most likely, did not contribute much, if anything, to either species gaining supremacy over the other as in competing for food and space. Where the two species have come together again at the split in the ring, they seem to intermingle quite well.

There are, however, other examples of mutations which may be considered beneficial. This kind of mutation seems to be confined to relatively small organisms with huge populations and short reproductive times. This group includes bacteria, insects, and plants, which respectively become immune to antibiotics, insecticides, and herbicides. These organisms apparently have undergone random mutations which turned out to be beneficial for their survival. However, none of these organisms have been known to advance to a different biological class. Each example remains confined in its original biological classification.

These examples of speciation, supposedly driven by non-silent mutations occurring in the relatively recent past, may be considered by some evolutionists as evidence of possible ancient stepping- stones that evolutionary processes randomly used in eons gone by to produce not just two new species but genus, families, orders, classes, phyla, and kingdoms. However, using an extrapolation of this magnitude as evidence of evolution's ability to produce the huge repertoire of biota in the ancient past is a step into chaos. Without memory, wrong combinations of genetic material produced by random mutations can be repeated an infinite number of times! Without intelligence or memory to recognize repeated errors, very minimal progress if any can be expected. Because evolution has no goal the proposed evolutionary progress would be helter-skelter at best. Pure random dumb luck is not sufficient to produce the plethora of biota on this planet in the time available. Greenish warblers are confined to greenish warblers. Bacteria, insects, and plants that apparently have mutated beneficially have remained in the same biological classification. Even though minor chemical changes may have occurred in their respective genomes, no significant biological advances have been observed beyond the species level.

Biological evolution, which demonstrates changes in biota at the species level is good science. However, any postulated changes beyond the species level ascending to order, class, phylum, or kingdom is an extrapolation into chaos.

Darwin could not know the limitations on his theory imposed by the huge amount of additional genetic information needed to change the genome of one class of biota into another because these discoveries were made many years later. He, therefore, extrapolated in his mind that there could be no limit to the beneficial changes that could accumulate when filtered by natural selection as thousands, or even millions, of generations came and went. Darwin thought that, given enough time, one species would change into something different, or even divide itself into two, which could eventually lead to two or more different taxonomic classes. He believed that all species were in a state of flux. After the limitations previously mentioned were discovered, scientists began to realize that thousands or even millions of beneficial mutations are necessary to add the required new DNA needed to gradually change a prokaryote into a eukaryote, over eons of time. Darwin's thinking about evolving one class into another, or even one kingdom into another, is best exemplified by a quote from *The Origin* where he says, "*By the theory of natural selection all living species have been connected with the parent-species of each genus, by differences not greater than we see between the varieties of the same species at the present day; and these parent-species, now generally extinct, have in their turn been similarly connected with more ancient species; and so on backwards, always converging to the common ancestor of each great class. So that the number of intermediate and transitional links, between all living and extinct species, must have been inconceivably great. But assuredly, if this theory be true, such have lived upon this earth." (The Origin of Species,* chapter titled "Imperfection of Geological Record.")

The third prediction located in chapter 6 of *The Origin*, Darwin states, "*If it could be demonstrated that any complex organ existed, which could not possibly have been formed by numerous, successive, slight modifications, my theory would absolutely break down.*" We can be sure that he was thinking of such organs as the eye, ear, or the liver. He did not know about complicated chemical processes such as the blood-clotting mechanism that forms part of the organ of the blood. It may come as a surprise that blood can be considered as an organ because of the many different kinds of cells working together like tissues in concert with chemical reactions residing in this red liquid. They all work together like an organ.

If it could be proven, as Darwin stated in *The Origin* that complex organs could not evolve by slight modifications over eons of time, according to him, his theory would absolutely break down. The test that Darwin proposed must pass muster of the confining boundaries imposed by irreducible complexity.

The confining boundaries of irreducible complexity can be easily demonstrated by a mousetrap. This illustration is borrowed from the book *Darwin's Black Box* by Michael J. Behe, and is used with his permission. (Michael Behe, at the time of the writing of his book, was associate professor of biochemistry at Lehigh University in Pennsylvania). A mousetrap consists of a wood platform, four staples, a bait holder, a spring, a hammer, and a holding bar. Remove any one of these pieces and the trap ceases to function. In other words, for the mousetrap to work, all the pieces must be in their respective places at the same time for the trap to function. At the level of irreducible complexity, function stops if one piece is missing. The mousetrap is irreducibly complex as it exists. Now, in this example of the mousetrap, if some irresistible bait is placed on the bait holder, it will increase the function of the mousetrap. On the other hand, remove any one piece of the trap and function ceases. This is true regardless of how many mice there are or how attractive the bait is. When acting in concert with all components of the mousetrap in place, function exists, and mice can be caught. However, remember that each individual piece of the trap by itself is functionless. The confining parameters of irreducible complexity can also apply to many biological processes as well, not the least of which is the blood coagulation cascade.

A close look at the diagram of the blood coagulation cascade as shown in Figure 11-1 will demonstrate the presence of more than 15 proteins that must react in concert to stop the bleeding of a simple cut on your finger. This cascade has many checks and balances inherent in it to prevent over-reaction or under-reaction. If over-reaction occurs, blood clots will form inside arteries which can cause strokes or heart attacks, to mention only two complications.

If on the other hand, the cascade under reacts, hemorrhage will occur with minor trauma such as a small cut. Hemophilia is an example of the latter, which is an inherited condition found mostly in male children. In hemophilia one part of the cascade is missing or malfunctions causing unwarranted hemorrhages. To correct the problem hemophiliacs must be given blood products obtained from other people, which contain the missing pieces. This must be done repeatedly. Hemophilia originated from a non-lethal detrimental mutation in the DNA of the blood coagulation cascade. The mutation occurred many generations ago but under certain conditions can be passed on to descendants of people who are carriers of the mutation, but who do not have the disease. Hemophilia shows that when one piece of the blood coagulation cascade is missing, function ceases. The cascade is irreducibly complex as it exists.

This cascade, which really is a system of checks and balances, is mostly mediated by the many proteins that participate in it. We have already learned how complicated one protein is, yet each of these proteins is functionless by itself, just like the individual pieces of the mousetrap. Natural selection cannot be involved in preserving or rejecting any of these proteins because each is functionless and therefore there is nothing to trigger its choosing process. Each must be present simultaneously for function to exist.

Evolutionary theory, as envisioned by Darwin, supposedly is empowered by natural selection to improve or perfect biological structures or processes that already exist. Natural selection must perform this feat without the benefit of memory, goal, or intelligent guidance. Evolutionists believe that all biological processes, including the blood coagulation process, have arisen by very small increments. Since almost all individual components of the blood coagulation cascade are functionless by themselves, it would be impossible for natural selection to operate in this situation. At some point in the distant past the blood coagulation cascade had to appear with all of its complicated parts present simultaneously in order for function to exist. According to evolutionary theory, when the first animal with a circulatory system came on the evolutionary scene, it had to have had a cascade already in place in order for that first primitive animal to survive. Although the individual cascades of animals have probably not been studied in as depth as this process has been studied in humans, it is obvious that they exist with similar checks and balances as those that exist in humans. Mice, minks, birds, and alligators along with monkeys, rats, whales, and snakes can survive injuries where blood vessels have been opened by some type of trauma. All of these examples, and more, must each have the blood coagulation cascade with checks and balances to prevent unneeded clots or unwanted hemorrhages.

The blood clotting cascade could not begin simple and become more complex with the passage of eons of time through slight modifications because all the components had to be present simultaneously. There is no way that evolution could produce all the complicated parts simultaneously. The very rare beneficial mutations proposed by evolutionary theory can theoretically only improve something that already exists. Beneficial mutations cannot produce a whole group of necessary proteins simultaneously from those that didn't exist before. Anything less than this set of requirements would cause natural selection to discard it to the theoretical trial and error trash heaps of evolutionary theory.

THE BLOOD COAGULATION CASCADE. PROTEINS WHOSE NAMES ARE SHOWN IN NORMAL TYPE FACE ARE INVOLVED IN PROMOTING CLOT FORMATION: PROTEINS WHOSE NAMES ARE ITALICIZED ARE INVOLVED IN THE PREVENTION. LOCALIZATION. OR PREMOVAL OF BLOOD CLOTS. aRROWS ENDING IN A BAR INDICATE PROTEINS ACTING TO PREVENT. LOCALIZE. OR REMOVE BLOOD CLOTS.

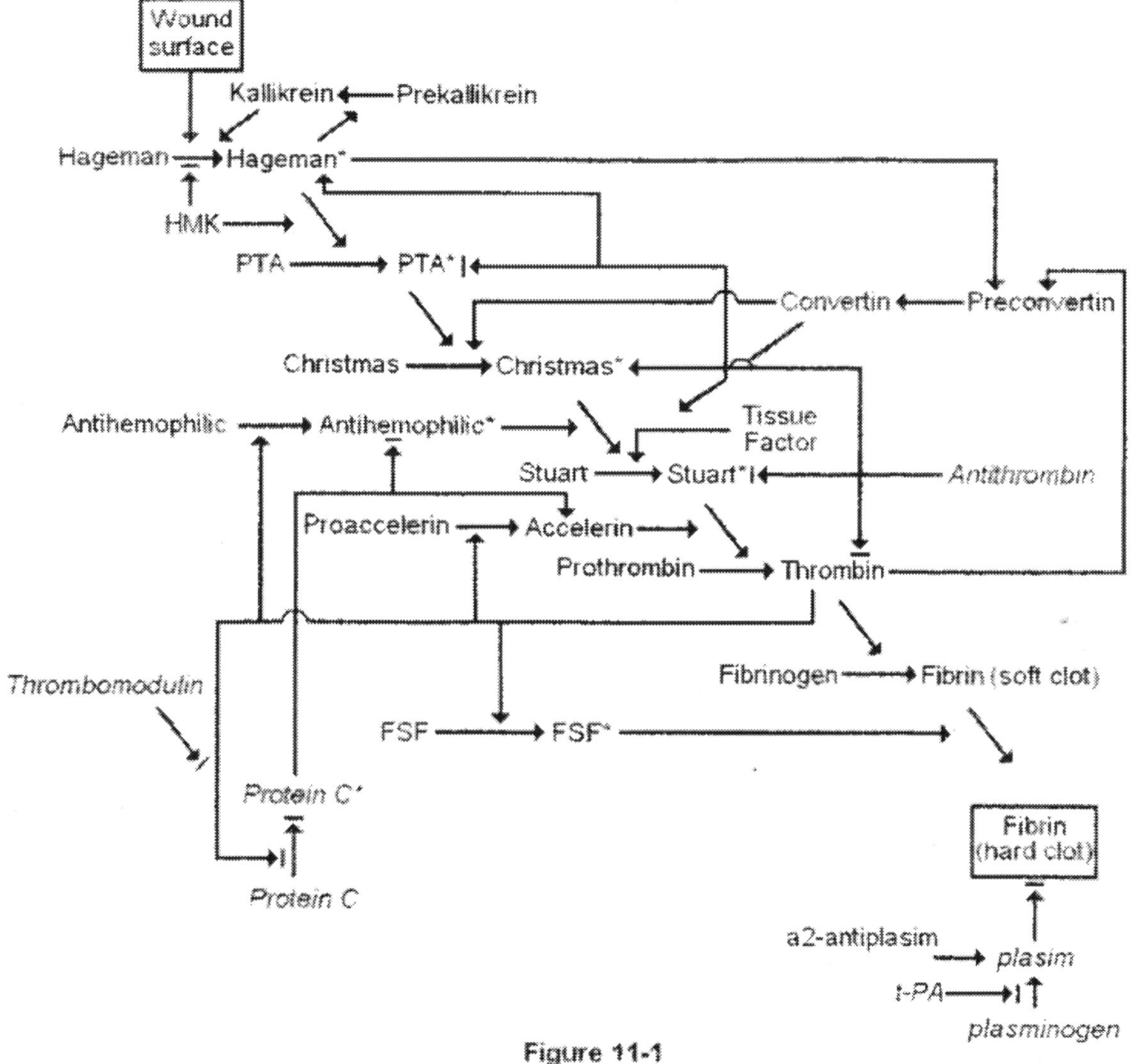

Figure 11-1

Used by permission of Michael Behe

It is impossible to conceive any stepwise method for this very necessary vital mechanism to have formed by numerous, successive, slight modifications as evolutionary theory would have us believe. There are many other biological examples of irreducible complexity that could be cited, each of which provides one more negative circumstantial evidence against Darwin's theory. Each could not start simple and become more complex by chance mutations accepted or rejected by natural selection. They had to start complex at the outset and remain that way generation after generation for function to exist. To add even more insult to evolutionary theory, how did all the information for making all the proteins needed for the blood-clotting mechanism first get stored in the DNA?

With the advent of sexual reproduction, especially in multi-celled eukaryotes, the question immediately comes to mind, how could evolution produce sex? A better way of stating this would be how could evolution supposedly empowered by the forces of natural selection produce two sexes, each with different reproductive organs and each with one complete chromosome difference? This question becomes even more provocative when it is remembered that Darwin's theory implies that all biota has ultimately been derived from simpler organisms that reproduce asexually. In other words, how could asexual organisms be progenitors of organisms with two sexes? This question still hounds evolutionary scientists as noted by the first sentence in an article titled "Sex, Sunflowers, and Speciation." "Why sex evolved and is maintained in most living organisms, remains a key question in evolutionary biology." See *Science*, 29 August, 2003, Vol. 301, No. 5637, p. 1189. This problem was on Darwin's mind and caused him to study barnacles for eight years. This was because he discovered that some barnacles are hermaphrodites; that is, both sexes exist in the same individual. Stated differently, a hermaphroditic barnacle can fertilize its own eggs. The second level of barnacles is hermaphroditic and also can mate with small, free-swimming males. The third group reproduced strictly by the mating of female barnacles with small, free-swimming males, which take up permanent residence in the larger female, sometimes two males at a time. The sole purpose of these resident males is to fertilize the female eggs. Darwin thought, through his study of barnacles, that he had found the answer to how evolution had produced the two sexes. First, there was the hermaphrodite, which had sex with itself. Next was the hermaphrodite that had sex with itself and with males. Last, the barnacles that reproduced only by the mating of male with female, in other words, true heterosexual reproduction.[8]

Maybe it didn't occur to Darwin, but the example he cited from the barnacles as an explanation for the origin of sex leaves more questions unanswered, than it answers. For instance, from what did the free swimming male barnacle descend modified, especially when it is noted that the male barnacles that Darwin studied had not one, but two penises?[9] Of course he knew nothing about chromosomes, but now is it not fair to ask where did the Y chromosome come from?[10] From what did these organisms descend modified? The origin of male and female organisms from asexual predecessors remains a mystery enclosed in another confining boundary.

It must be emphasized again that the pepper-moth experiment did not produce a new species of moth, but supposedly proved that Darwin's notion of natural selection is at work in the wild and, as previously stated, the moths stayed moths. The same notion applies to various birds colonizing the archipelagos of Galapagos and Hawaii and to the cichlid fish of Lake Victoria. Natural selection cannot be proved as the cause of these two examples of speciation. There is no living proof or fossil evidence that one class of organism has ever evolved into another class, such as a fish evolving into an amphibian. We have already noted improbabilities of producing a prokaryote from abiotic material. After that, consideration must be given to the possibilities for evolution to produce a single-celled eukaryote from a prokaryote. Then, we must consider the possibilities of changing a single-celled eukaryote into a multi-celled eukaryote. These three different processes pose huge summits over which evolutionary theory must climb if its truthfulness is to be considered credible. These three summits are actually overhanging cliffs with no gradual inclines on the other side. They pose much greater difficulties for evolutionary theory to explain than changing one class of biota into another as in changing a fish into a frog. These three confining boundaries actually produce a cage that forms a prison from which Darwin's theory can't escape.

SUMMARY

1. Because DNA and its genetic code define the transmission of genetic characteristics from one generation to the next in all life forms, this fact suggests that all living organisms could be related and have descended modified from one primordial cell. This is a form of positive circumstantial evidence.
2. Darwin's theory is based on positive circumstantial evidence.
3. Circumstantial evidence comes in two forms: positive and negative. Positive circumstantial evidenc shows that something could have happened in a certain way but never that it did. Negative circumstantial evidence is more powerful. It can show that something could not, or did not, happen in a certain way.
4. The theory of evolution postulates that biota started simple and became more complex over eons of time, gradually dividing itself into the myriads of species contained in five kingdoms, through the driving force of natural selection.
5. For evolutionary theory to be proven true, there must be many elements of positive circumstantial evidence and no negative circumstantial evidence.
6. Does natural selection work in the wild benefiting one group of organisms and acting in a detrimental way to others? Supposedly the peppered-moth experiment proved that it does work both ways in the wild.
7. Darwin's theory had to wait for nearly one hundred years before some evidence of natural selection in the wild could be documented. This was supposedly done by Kettlewell's peppered-moth experiment, which now has come under critical review.
8. Even if the results of the peppered-moth experiment ultimately are proved true and even though natural selection in the wild is present, these two processes do not demonstrate that new species arise by this method.
9. New species of birds and fish have arisen in the wild, but natural selection cannot be proved as the cause. Birds stayed birds, and fish stayed fish. No direct evidence exists showing how one class of organism has changed into another class, even over eons of time.

10. Geographic isolation seems to have contributed to the development of new species of birds which arose from colonies that originally populated the archipelagos of the Galapagos and Hawaii. However, birds stayed birds, very similar to the original colonizers of these islands. In some cases, they have not become so widely separated that they cannot interbreed. One new species of greenish warbler has been identified in Siberia, resulting from geographic separation. These birds apparently have become reproductively isolated from the species to which they seem to be related.
11. Fish in ancient lakes also have been known to develop new species. But again, natural selection cannot be proven to be the cause of the new species.
12. Beneficial mutations seemingly have been observed in bacteria, insects, and plants that respectively have become immune to antibiotics, insecticides, and herbicides.
13. Irreducible complexity exists when removal of any one part of a system composed of many parts causes the system to cease to function. A mouse trap is a classic example. Remove any part and function ceases.
14. The same idea applies to many biological systems. When one part is removed, function ceases.
15. There are many complicated parts composing the blood-clotting mechanisms. Clotting is prohibited when any one of these complicated parts is missing.
16. Hemophilia is an example of a deleterious mutation confined almost exclusively to male children who are born with one complicated part missing.
17. Because survival of the fittest, or natural selection, cannot detect something that does not function, it would be impossible for it to perfect the blood-clotting mechanism one part at a time. This is because not one of the individual parts has any function by itself. In order for the blood-clotting mechanism to function, all the complicated parts had to be put together simultaneously. The whole mechanism of blood clotting is irreducibly complex as it is. When any one part is missing, function ceases. Function begins when all parts are present and act in concert. Therefore, evolution could not produce this mechanism one part at a time by slowly perfecting each individual part. This is another example of negative circumstantial evidence.
18. Evolution has three huge overhanging cliffs over which it must climb if we are to believe it true:
    a. The formation of the first prokaryote from the "primordial soup"
    b. The formation of the first single-celled eukaryote from a prokaryote
    c. The formation of the many kinds of multi-celled eukaryotes from the single-celled Eukaryotes. Each of these three are unrelenting confining impasses.

Chapter 12

The Appearance of Single-Celled Eukaryotes

The unrelenting frustrations that many scientists have encountered in the presupposed chemical evolution of prokaryotes arising spontaneously from the primordial soup have been discussed in chapter 9 and in addition, other confining parameters were described in Chapter 11. However, if the first prokaryote could have arisen spontaneously, this would be the boundary line between chemical evolution and biological evolution, because the latter requires at least one living organism present at the outset for the evolving changes to begin that Darwin envisioned. In the last paragraph of *The Origin of Species* he postulated that life began with a "few forms" or even "one." Next, consideration must be given to the difficulties encountered for evolution to evolve single-celled eukaryotes from prokaryotes and in so doing totally ignore the extreme improbabilities of the origin of the first prokaryote arising spontaneously from the primordial soup as explained in chapter 9. Without proof of its origin, at least one prokaryote will be assumed present for biological evolution to begin. An article by a Nobel Prize winner, C. de Duve, in the April 1996 *Scientific American,* (Vol. 274, pp. 50-57) endeavors to explain this supposed evolutionary event. At the outset, de Duve enumerates the main differences between prokaryotes and single-celled eukaryotes.

1. Eukaryotic cells are as much as ten thousand times larger in volume than prokaryotes. (Some prokaryotes are spherical as are some single-celled eukaryotes and some of each are not. However, for the sake of getting a better understanding of the differences in size between these two life forms, think of each cell type as spherical. With only about a 26 percent increase in the radius of a sphere, the volume doubles. However, with an increase in volume of 10,000 times the radius of a sphere would have to increase about 14 times. Even though many times larger than prokaryotes, almost all single-celled eukaryotes are microscopic in size. This gives us a better perspective of just how small they both are. Until the invention of the electron microscope, very little was known about their respective internal structures.)
2. Eukaryotes have most of their genetic material housed in the nucleus in individual pairs of homologous chromosomes, in contrast to the single circular chromosome of bacteria or prokaryotes, which have no nuclei.

3. Because the eukaryotic cell is so much larger than the prokaryote, the cytoplasm surrounding the nucleus is divided by various partitions "into an elaborate network of components that fulfill a host of functions." (page 50)
4. There are internal skeletal structures that provide support to the eukaryotic cell contrasted with a prokaryote, which has a tough cell wall on the outside to support it.
5. The size of the eukaryotic cell is so big that it requires a transport system to move the various products of metabolism from one site in the cell to another. This includes new parts manufactured and waste products to be excreted. Tiny "molecular motors" carry this out as contrasted with the prokaryotic cells that depend upon diffusion and concentration gradients to perform these tasks.
6. Eukaryotic cells have specialized organelles, which may number in the thousands, each of which may be the size of a prokaryote.

Keep in mind the various organelles just mentioned. de Duve points out (page 50) that besides the nucleus, three of these are also very important; "peroxisomes (which serve assorted metabolic functions), mitochondria (the power factories of cell)," and "plastids (the sites of photosynthesis)" in algae and plant cells.

de Duve postulated three assumptions that might represent how the evolution of eukaryotes from prokaryotes might have occurred.

1. This first proto-eukaryote developed from a prokaryotic cell that "fed on debris and discharges of other organisms." (page52) This type of prokaryote does exist and is a heterotroph. de Duve theorized that this heterotroph lived "in mixed prokaryotic colonies" (page 52) that have been found fossilized in sedimentary rocks called stromatolites.
2. de Duve hypothesized that this proto-eukaryotic cell digested its food as living heterotrophic prokaryotes do (outside its cell) by secreting enzymes to do the digestion before the food was absorbed.
3. This hypothetical ancient organism, "…lost its ability to manufacture a cell wall, the rigid shell that surrounds most prokaryotes, and provides them with structural support and protection against injury." (page 53)

de Duve pointed out that living naked forms of this type of prokaryote exist today.

At first glance these assumptions do not seem preposterous because they are based on present-day living organisms that display these particular individual characteristics. As this proposed proto-eukaryote increased in size, all structures needed in the "modern eukaryote" would have been required for this cell to function. de Duve admits to this, in a quote: "*Modern eukaryotic cells are reinforced by fibrous and tubular structures, often associated with tiny motor systems that allow the cells to move around and power their internal traffic. No counterpart of the many proteins that make up these systems is found in prokaryotes. Thus, the development of the cytoskeletal system must have required a large number of authentic innovations. Nothing is known about these key evolutionary events, except that they most likely went together with cell enlargement and membrane expansion, often in pacesetting fashion.*" (page 55)

THE SECOND LEAP OF LIFE: PROKARYOTES TO EUKARYOTES

When Darwin made his theory public in 1859, virtually nothing was known about prokaryotes and single-celled eukaryotes; and certainly nothing was known about the complicated differences between these two life forms. If evolution is responsible for the slow, step by step transition from the simpler prokaryote to the more complex, single-celled eukaryote, it had to replace the circular chromosome of the prokaryotes with the paired homologous chromosomes of the eukaryotes. These paired chromosomes, instead of residing naked in the cytoplasm, as is the case with the prokaryote's single, circular chromosome, these reside in an organelle known as the nucleus. Except for the mitochondria and/or chloroplasts, which contain their own DNA, that nucleus had to contain in its DNA all the information needed, not only for making the nucleus itself, but also the patterns for making all the other kinds of organelles plus the hundreds of other structures needed for eukaryotic metabolism. The aforementioned paired homologous chromosomes collectively had to contain hundreds of new genes, each of which would be composed of at least one hundred, or even up to one thousand or more, perfectly sequenced codons. Each of these in turn would be composed of three perfectly sequenced base pairs. If evolution, as the model of the Neo-Darwinian synthesis envisions, is responsible for the diversity of biota on this planet, which started with a minimum of at least one simple life form, most likely a prokaryote, then the genetic code must have been present in that first organism for replication to comence and for mutations to occur.

There are millions of base pair differences between these two life forms. These differences include not only an increased amount of new genetic material but also rearrangement of old genetic material now located in the single-celled eukaryote. This genetic material was supposed to have been inherited from its postulated predecessor, the prokaryote. This is not an exaggeration; there are more than seven million more base pairs in the genome of one so-called simple yeast Saccharomyees cerevisiae (a single-celled eukaryote) than occur in the genome of one so-called simple bacterium E. coli (a prokaryote). These millions of changes, evolutionary theorists tell us, must result from beneficial mutations, which had to occur in perfect sequence to accomplish this huge leap of life. Even de Duve points out that "nothing is known about these key evolutionary events, except that they most likely went together with cell enlargement and membrane expansion, often in pacesetting fashions." (page 55) It is obvious that huge quantities of genetic material must be added and old genetic material modified, for prokaryotes (the kingdom of Moneria) to evolve into single-celled eukaryotes (the kingdom of Protista). This information was not available to Darwin in 1859.

For the trial and error method of Darwin's theory to produce by chance the huge number of necessary beneficial mutations with perfect placement, certainly stretches the imagination, to say the least. These changes are even more incredible when we consider that evolution is blind and has no memory, intelligence, or goal. Someone might say that the goal of natural selection is survival. The theory of natural selection, however, allows beneficial mutations to survive and deleterious ones to be discarded. A goal is always future oriented, whereas survival is always about the here and now. Natural selection has no intelligent future goal. Without a memory, the more frequently occurring deleterious mutations would be repeated over and over again, each one of which natural selection would have to reject, thus producing multiple and inevitable dead ends. They therefore would have a tendency to erase any progress that may have been made by the theoretically proposed very rarely occurring beneficial mutations.

The late Ernst Mayr on page 149 of his book *This is Biology* states that life appeared "(about 3.8 billion years ago) until about 1.8 billion years ago, only prokaryotes existed." "Around 1.8 billion years ago the first one-celled eukaryotes originated, characterized by a membrane-closed nucleus with discrete chromosomes and by the possession of various cellular organelles." This means that for about the first two billion years after life appeared on planet Earth, it was populated only with prokaryotes.

It must be pointed out that origin-of-life paleontologists estimate that it took about 700 million years for life to appear on earth after it had cooled enough for life, as we know it, to exist. However, if evolution is responsible for the transition from prokaryote to eukaryote, it is strange that it would take two billion years for it to accomplish this feat. Theoretically, it should have taken less time because the transition from prokaryote to the single-celled eukaryote started with something already alive and supposedly ends with something alive, whereas the former started with non-living chemicals and supposedly ends with something alive. To summarize, biological evolution has two advantages over chemical evolution. 1) Biological evolution starts with a living organism, which in theory, it must change through millions of intermediate life forms into a different kind of living organism more complicated than the first. 2) It also has a method of choosing that is called natural selection, which chemical evolution did not have. Remember, in the chapter where chemical evolution was discussed, natural selection could not operate in this situation because it can make selections only between two living organisms, one of which competes better than another for food and space. Therefore, selection is based on differences in survivability and/or reproductive success. It cannot make choices between non-living organic molecules, because they are all functionless outside of a cell in a large open system such as a lake or ocean. Functionless chemicals cannot compete for survival. Although some organic molecules are more stable than others so some would have longer "half-lives" than the others, they cannot compete for survival. Concentration gradients can allow some chemical competition between various molecules, but dilution in the ocean would prevent this. However, after the first life forms came into being, natural selection theoretically could evaluate conditions that might allow beneficial mutations to survive and deleterious mutations to be discarded. This is because an organism containing a beneficial mutation theoretically would function better and compete more effectively for food and space than an organism that did not possess one. Therefore, they would leave more progeny like themselves.

Evolutionary theorists postulate that life arose spontaneously at some place on the early Earth in an environment rich in organic molecules and an energy source. Deep ocean vents where heated water brings up a rich mixture of chemicals from deep inside the earth are considered by some evolutionary theorists to be ideal locations for simple first-life forms (prokaryotes) to have arisen spontaneously. Other evolutionary theorists believe that the first life form arose near the surface in water pregnant with many carbon molecules and energy. In chapter 9, the unlikelihood for such an event to occur was discussed. Regardless of the extreme unlikelihood of such an event, for the sake of argument let us suppose that those extremely improbable odds were somehow overcome and the first life form LUCA (last universal common ancestor), most likely a prokaryote, "magically" arose from the hostile abiotic chaos. This first life form had to be able to grow, metabolize chemicals for energy, and genetically replicate.

Regardless of its reproductive time, whether minutes, hours, days, or weeks, there would have been no competition at first from other organisms for food and space in this free-lunch environment. Therefore, we can assume that the first prokaryote could multiply unimpeded until the population became large enough for competition to begin and for natural selection to start sorting among variations that mutations may have presented to it.

Remember that prokaryotes reproduce by simple cell division, which means that a given individual does not have two parent's; it simply has an identical twin both of these resulted from the division of one previous mother cell. Because of this, until recently it had been proposed that prokaryotes enjoy a form of immortality, where aging did not exist. In other words, unless a given prokaryotic cell was killed by some outside causes, such as excessive heat, lack of food, antibiotics, etc., it would not age because it simply divided itself into two new cells. This concept changed by recent research reported in the February 4, 2005 issue of *Science* (page 656) where it was shown that certain E. coli cells do age. If the aging factor and losses caused by external causes are ignored, the equation for the unimpeded rate of multiplication of prokaryotes is $P = 2^n$, where P equals the number of individual prokaryotes in the population at any given reproductive cycle, and n equals the number of individual reproductive cycles. This is a form of exponential growth by which a bacterial population can expand very quickly, as you will soon see.

For the sake of argument we will imagine that the first life form, a prokaryote, appeared spontaneously from an abiotic environment rich in organic molecules. Having started with one prokaryote as our beginning limit, the probability for this first prokaryote to eventually evolve into a single-celled eukaryote by the end of two billion years will be estimated. From this point on the use of the words prokaryote and bacterium will be used interchangeably. However, there are some bacteria that can reproduce themselves in 20 minutes yielding three generations per hour. This is exponential reproduction. Roger Y. Stanier et al. in the book *The Microbial World*, (page 185) points out that "Microbial populations seldom maintain exponential growth at high rates for long. The reason is obvious if one considers the consequences of exponential growth. After 48 hours of exponential growth, a single bacterium weighing about 10^{-12} grams with a doubling time of 20 minutes would produce a progeny weighing 2.2×10^{31}grams, or roughly 4,000 times the weight of the earth." The mathematics for this is presented on the next page. You do not need to do the math but simply proceed to the conclusion, shown in bold type.

48 hours x 3 doubling times per hour = 144 doublings
$P = 2^n$; where n = 144 Ln 2 = 0.693147181.
This number multiplied by 144 = 99.813194. The Anti Ln = $2.23007452 \times 10^{43}$ bacteria after 48 hours.
This number ($2.23007452 \times 10^{43}$) multiplied by the weight of one bacterium which is 10^{-12} = 2.23×10^{31} grams, which is approximately equal to 4,000 times the weight of the earth. Obviously there could not be a colony of bacteria with that much weight.

By dividing $2.23 \times 10^{31} \div 4 \times 10^3 = 5.575 \times 10^{28}$ grams equals the approximate weight of one earth. Therefore, the total bacterial population, though potentially very large, must have a weight much less than one earth. Divide this number, 5.575×10^{28} by the weight of one bacterium 10^{-12} = 5.575×10^{40}, which equals the number of bacteria weighing approximately equal to the earth. This means that exponential growth of any bacteria cannot continue for very long. This includes our first proposed bacterium, the primordial prokaryote.

Stanier et al. explain in their aforementioned book (page 185) that bacterial growth, or reproduction, proceeds, starting first with a lag phase where they reproduce slowly. This is followed by the exponential growth phase where the bacterial population increases very rapidly ($P = 2^n$). For reasons alluded to above, this rapid growth cannot maintain itself for long. The exponential growth phase is followed by a stationary phase in which reproduction is almost at a standstill. In turn this is followed by a death phase. Stanier et al. also tell us that the rate at which bacteria die during the death phase is also exponential. The death phase results from several causes: lack of nutrients, build up of toxic metabolic waste products, and the resulting depletion of cellular reserves.

It can be assumed that the first primordial prokaryote or bacterium, after going through the lag phase, entered the exponential growth phase. It is hard to postulate a situation around a deep hot water ocean vent or a similar situation near the surface, where nutrients could be exhausted and/or waste products build up to the point that the stationary phase, or death phase, ensues. Remember, in this primordial situation, there would be no competition for food or space from other forms of biota. However, it would be easy to visualize that as the bacterial population grew, some of the population would have been wafted away by the currents around the vents or streams near the surface so that many would have entered the lag phase. Following this, it is easy to understand that a few bacteria by chance were carried to another place where they could again reproduce exponentially for a few hours. The bacterial population in these reproductive cycles could be repeated millions of times forming graphically a saw-toothed appearance, as they repeatedly multiplied and died in their respective population cycles and spread away from their point of origin.

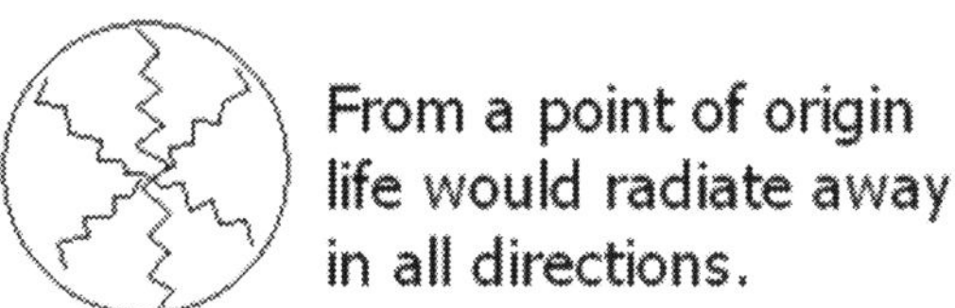

From a point of origin
life would radiate away
in all directions.

Figure 12-1

This way with minor variations, prokaryotes according to evolutionary theory, could have been able to populate our entire planet over a two billion year time frame. We will now be ready to explore the probabilities of evolving a single-celled eukaryote from a prokaryote starting with one bacterium and 2 billion years to do the job.

As bacteria spread from their point or points of origin, they eventually encompassed the entire earth. They have managed to occupy more ecological niches than any other group of living organisms. Bacteria have been discovered in rocks removed by oil well bits 1000 feet beneath the surface of the earth. They live in the ocean, in soil, and in the air. In fact the entire planet can be thought of as a bacterial eco-system. From the mathematics of exponential replication just described by Stainer et al. it is obvious that if all the bacteria in the entire world suddenly entered the exponential replication phase, the planet would be covered in just a few hours by a thick layer of bacteria. The main limiting factor for this to occur is the lack of sufficient food. From this, we can derive that almost all bacteria exist in the lag phase, where replication is very slow. However, if a bacterium happens to land in the friendly environment, where is located the correct temperature and an abundance of food, the lag phase suddenly gives way to the exponential replication phase. But alas, due to insufficient food, this situation can continue for only a few hours at best. Here again, it becomes clear that the bacterial population actually can be said to be in equilibrium with the food available on planet Earth, otherwise we would be inundated by them. When the total bacterial population of planet Earth is considered, it becomes obvious that there are relatively few colonies of bacteria that are replicating exponentially at any given time, outside the guts of warm-blooded animals.

Paleontologists are usually thought of as scientists who study fossils, all of which have long since been dead. However, a new twist in paleontology has emerged among molecular biologists, who study metabolic pathways that seem to have survived the rigors of more than three billion years of supposed evolutionary pressures. Chemicals composing the citric acid cycle and RNA seem to have undergone little change from multi-celled sexually reproducing eukaryotes down to single-celled eukaryotes and on down to prokaryotes. These two are thought to represent chemical processes that originated in the most "primitive" life forms and have been passed on through eons of time to living forms existing today. This is a vicarious way of studying the dead among the living. Scientists who conduct research along these lines are called molecular phylogenists. Professor Robert M. Hazen, Ph.D. (as noted in lecture 15 in his course Origins of Life by the Teaching Company), explained that the University of Illinois geneticist, Carl Woese is the acknowledged pioneer of molecular phylogeny. The evolutionary variations involving these two chemicals and their pathways have not been as great as might be expected.

Evolutionary advancement is dependent upon random timing and placement of mutation changes in the DNA that are presented to natural selection for acceptance or rejection. In fact, the most important prerequisite to theoretical evolutionary advancement is the mutations themselves, which can change DNA instructions permanently. There is no question that silent, neutral, and various kinds of deleterious mutations exist, but for natural selection to produce gradual perfection in any given species, beneficial mutations must also exist. The existence of beneficial mutations then is the most critical basic requirement for the advancement of evolutionary theory, but their existence may be an assumption!

One bit of evidence proposed for existence of beneficial mutations in prokaryotes is the development of resistance to antibiotics in some bacteria. A classic example is the methacillin resistant Staphylococcus aureus, or MRSA, as they are called. It appears that these bacteria have developed a beneficial mutation (or mutations) that give them the ability to resist the devastating effects on themselves of most antibiotics. Of course a mutation like this would be beneficial to the bacteria but not to a person suffering from an infection caused by these bacteria. A mutation beneficial to this bacteria's existence, making it able to survive in an environment hostile to other bacteria that do not contain this property of antibiotic resistance, would give it a definite advantage that natural selection would allow to survive. This group of bacteria will therefore leave more progeny with the same advantage. However, the ability of certain bacteria to resist antibiotics may not represent a true mutation in many cases at all. Many plasmids, which have extra chromosomal segments of DNA or RNA similar to viruses, reside like an intracellular parasite inside many bacterial host cells. However, unlike viruses they do not form infectious particles. Often their genetic messages encode enzymes capable of providing antibiotic resistance to the host cells in which they reside, as a kind of symbiotic relationship. There may be many plasmids residing in a single host bacterium, so that when the host divides some plasmids will be found in each daughter cell. If this line of bacteria becomes involved in an infection they will thrive even in the presence of many different kinds of antibiotics. Therefore antibiotic resistance may not always be the result of a beneficial mutation after all.

In addition to what has just been noted in the preceding paragraph, some bacteria have a built-in method in their metabolism for very rapidly mutating DNA to produce a host of new proteins, which result in self-defense against antibiotics. This defense mechanism is called the SOS response because it occurs rapidly in some bacteria that become exposed to certain antibiotics, which otherwise would cause their demise. This phenomenon has been known by molecular biologists since 1970. Not all bacteria contain this defense mechanism in their metabolism. However, one bacterium that does is Escherichia coli. The April, 2006, Scientific American describes what happens when E. coli bacteria are exposed to the antibiotic Cipro. This antibiotic works by attacking gyrase, an enzyme needed by E. coli for its normal DNA replication. This triggers the SOS response, which is mediated by switching on "genes whose protein products precipitate a spate of mutations that occurs 10,000 times faster than those arising during normal cell replication." One of these many new proteins resulting from this sudden burst, accidently turns out to be an antibody that can protect the bacterial protein gyrase from attack by Cipro.

Floyd E. Rosenberg Ph.D., a research chemist working at Scripps Research Institute in La Jolla California, became interested in this SOS process. He along with 19 other researchers and "coworkers Ryan T. Cirz, Jodie K.Chin and their collaborators at the University of Wisconsin-Madison," discovered how the SOS response is mediated by the bacterium E. coli. They found that a protein known as LexA, occurring naturally in these bacteria, acts as a repressor to the SOS response. The LexA protein must first be cleaved, which then allows three DNA polymerases to manufacture a barrage of mutated proteins. One of these proteins is able to protect gyrase, and thereby prevent Cipro from killing the bacterium in which this process occurs. Once a given bacterium becomes resistant to Cipro, all daughter cells descending from this original bacterium will also contain the defense mechanism. The above information was taken from Scientific American April 2006, Vol. 294, number 4, pages 81-83.

Because this metabolic reaction causes the production of many new mutated proteins, which in turn prevents the death of the bacterium in which it occurs, some scientists point to this phenomenon as evidence of fast track evolution. But isn't this reminiscent of what takes place in the B lymphocytes (B-cells) in the immune system of higher animals and humans? Through very rapid mutation rates (a million times faster than usual) gene segments in specific B-cells called hybridomas, are rearranged so they can produce millions of different proteins. One of these turns out to be a specific antibody needed by their immune systems to react with invading antigens such as viruses or bacteria (Kleinsmith, Lewis J., Kish, Valerie M. Principles of Cell and Molecular Biology page 726).

Perhaps, the rapid SOS response in E. coli bacteria is nothing more than a primitive and simpler kind of immune system found in higher eukaryotes and has nothing to do with evolution per se. An E. coli bacterium made resistant or immune to the presence of Cipro by the SOS response, did not evolve into something else, but remained an E. coli, now simply immune to Cipro. This is very similar to a human who remains human after being made immune to the chickenpox (varicella) virus by B cells, which produce via rapid mutation rates, a specific antibody (a protein) against the virus through multiple rearrangements of genes. In both cases a new protein was made, which performed a specific function. In humans a process that manufactures brand new proteins to ward off intruders is called an immune response, whereas in bacteria a similar process is called by some scientists as fast forward evolution. Perhaps the molecular phylogenists should look into this matter. Maybe the SOS response in bacteria is as ancient as RNA or the citric acid cycle. After all, bacteria are found in almost every location on planet Earth including rocks retrieved from oil well drill bits taken from 1000 feet underground. Even though bacteria are very prolific, they must also contain some defense mechanisms in order to avoid annihilation. The SOS response seems to be a defense that they can turn on when exposed to certain antibiotics.

We have seen in the two examples above, one in prokaryotes and the other in multi-celled sexually reproducing eukaryotes, that specific environmental conditions stimulate their respective metabolisms to very rapidly mutate genes to produce multiple proteins, some of which are beneficial to the organisms in which they occur. No change in identity was noted in either case. Both examples remained ensconced in their respective species. This was true even though the mutation rate for the SOS response in prokaryotes was 10,000 times faster than normal and the mutation rate in the immune response of eukaryotes was one million times the normal. But how about usual random uncontrolled mutations occurring at much slower rates; can mutations under these conditions cause advancement in the complexity of living organisms? That remains to be seen.

A quote from *Principles of Cell and Molecular Biology*, (Lewis Kleinsmith and Valorie Kish, page 92) is in order here to set the stage regarding evolutions proposed random progress. *"Although spontaneous mutations can play an important role in generating the genetic changes that are required in order for evolution to occur, the immediate effects of most mutations are deleterious, and hence an excessive number of mutations is undesirable."*

We have learned that the most complex prokaryote contains millions of base pairs less than the simplest single-celled eukaryote. Although logic cannot play a part in evolutionary processes, it seems reasonable to assume that the easiest, simplest, and shortest way for evolutionary advancement to occur would be to first have introns or junk DNA added to the prokaryotic genome, which later could be modified by mutations to form genes beneficial not only to the prokaryote in which they supposedly are occurring but also beneficial for the single-celled eukaryote, which might inherit them millions of years later. It is difficult to imagine how even one new gene out of about 1500 more needed by a single-celled eukaryote could be slowly added to the genome of a prokaryote without first having a large chunk of junk DNA added, on which beneficial mutations then could later work. Therefore, one of the first accidental tasks incumbent upon evolutionary processes must be the addition of a significant amount base pairs. If this occurred first, then these added base pairs probably would not disrupt the metabolism of a prokaryote as a deleterious mutation might do, because they would operate like junk DNA. But can any kind of mutation add significant amounts of DNA to the genome of a prokaryote? This is the next question that automatically comes to mind.

As stated above, the genome of a single-celled eukaryote contains many more base pairs and genes than a prokaryote. We must ask ourselves if there is any way to estimate the chances to evolve a single-celled eukaryote from a prokaryote by the random placement and random timing and sequence of mutations in the DNA of a prokaryote? We will start by checking out the frame shift mutation which either adds or subtracts one base pair from the genome of the affected cell. This type of mutation is especially deleterious because it can seriously damage an entire gene.

Stem and loop mutations are available for adding base pairs to our theoretical evolutionary process in the prokaryote but over time, they have a 50-50 chance of adding or subtracting genetic information in the genome in which they occur. Therefore they can be involved with changes in the genome but probably would not contribute very much by adding new DNA.

We have learned in chapter 7 that there is a type of mutation occurring in prokaryotes that can add genetic information to the genome. It is called a **transposable element**. Transposable elements come in two different categories, **insertion sequences** and **transposons**. Insertion sequences contain only a few thousand base pairs, where as transposons tend to be much larger and may contain numerous genes. Transposable element mutations act on the genome by removing a segment of DNA located in one place and reinserting it back into the genome in another place. Each type of transposable element mutation depends on a protein known as a **transposase,** which is encoded by a gene located in the DNA that is being transferred. Sometimes these mutations manifest themselves by replicating a segment of the DNA at a given location, which it transports and reinserts into a new location but leaves the original segment in its original place. This process could obviously add new base pairs to the genome where it occurs, which beneficial mutations theoretically could change later. Because these mutations can cause rapid change in the DNA of any organism in which they occur, it seems logical that evolution would progress faster if the gene for the transposase protein became available early on in any attempt to evolve a single-celled eukaryote from a primordial prokaryote. A quote from Kleinsmith and Kish, page 96, seems very appropriate at this juncture. "One of the most interesting properties exhibited by transposable element mutations is that in the process of moving from site to site, they influence the expression and organization of neighboring genes. Since transposable elements can migrate with considerable frequency, their behavior has the potential for causing rapid and dramatic genetic alterations. This phenomenon is now believed to be a major contributing factor to evolutionary change."

There is another protein known as an **endonuclease,** though not considered to be a mutation, is able to cut DNA at specific base pair sites. This protein could theoretically cut the circular chromosome of a prokaryote in several places, thus preparing the way for the evolutionary development of individual noncircular chromosomes found in the single celled eukaryote. This protein in conjunction with the transposable element mutations described above could theoretically pave the way for evolutionary processes to develop the individual homologous chromosome pairs found in all eukaryotes. First, the endonuclease could cut the circular DNA of the prokaryote producing non-paired chromosomes, which could be followed by a transposable element mutation which could reproduce one of the fragments making a homologous chromosome pair. (This is the way that we would look at the situation in retrospect but because evolution is blind, has no memory, intelligence or goal, logic could not enter into this proposed ancient accidental process). However, this does pose a possible evolutionary route for random mutations in a prokaryote to evolve into a single celled eukaryote. But before the above processes could take place, the primordial prokaryote must have in its genome the DNA for the proteins endonuclease and transposase.

Next, changes that can occur to any DNA molecule during a point mutation will be analyzed. Point mutations are also known as single base pair mutations because only one base pair in a given codon is mutated. In other words, one of the three base pairs that composes a given codon is accidentally exchanged for another. It must be emphasized that a point or base pair mutation does not add or subtract any base pairs from the total number of base pairs in a given DNA molecule. This kind of mutation only exchanges one base pair in a given codon to another. This means that the mutated codon is changed and with the next replication becomes permanent in the DNA, unless the same codon by chance becomes mutated again at a later time. A base pair mutation is probably one of the most common kinds of mutations, which has been estimated to occur about once in 10^9 base pair replications[1]. Recall from what you learned in chapter 7, that a point or base pair mutation requires that nucleotides in both the anti-sense and sense strand differ from the original DNA from which they originated. However, these nucleotides in a base pair mutation must be complementary. From this description of the point or base pair mutation it is easy to see that this type of mutation could be involved in changing a given codon one at a time but could not be involved in adding additional base pairs to the genome.

Besides mutations, there are three other ways that prokaryotes can add base pairs to their genome: transduction, conjugation, and transformation. Transduction occurs when either a bacterial phage or virus invades a bacterium's cell and introduces new DNA into the recipient's genome. This phenomenon is used in genetic engineering.

Conjugation occurs between two living bacteria. A tubular cytoplasmic bridge called a pilus is built between the two, through which genetic material passes from one bacterium to the other. This material may be a plasmid or a small circular DNA molecule that does not form part of the donor's genome. This circular bit of DNA sometimes becomes incorporated into the receiving bacterium's genome. The formation of the pilus is dependent upon a protein known as pilin.

Transformation is where some naked DNA, resulting from the lyses of another bacterium, (usually of the same species) is absorbed into a living bacterium. Sometimes this new DNA from a dead bacterium gets incorporated into the genome of the recipient cell, which is said to be cotransformed. Obviously, this would add new base pairs to the genome of a bacterium where the process of transformation occurred.

Each of the above situations increases the total number of base pairs located in a given recipient bacterium's genome. Even though the above phenomenon may explain the addition of base pairs to the bacterium's genome, and possibly the origin of introns in the DNA of the eukaryote, they don't really explain the breakup of the single circular chromosome of the prokaryote nor the origin of the homologous chromosomes nor the multiple organelles located in the single-celled eukaryote. However, even when additional base pairs are incorporated into the prokaryote's DNA, as with transformation, they must be modified later by mutations to form new genes. Most likely these mutations would be base pair or point mutations needed to form the new genes useful to a single-celled eukaryote that would inherit them later. In addition, the genes original to the prokaryote must also be modified to make them useful to the single-celled eukaryote. These changes would occur as Darwin envisioned by the shortest steps possible.

The genetic code has been described as being semi-conservative. The reason for this is that even though many base pair mutations are successful in striking many places in a given organism's DNA, in some instances no change in the code is observed following the mutation. In other words the original code is conserved some of the time, hence the term semi-conservative. The reason for this is that many base pair mutations are of the **silent** variety defined as follows. Silent point mutations are those mutations, which cause no change in the amino acid for which they code. (This excludes the situation where a given point mutation is mutated by the same base. Hence, if base A is mutated by another base A or base T is mutated by another base T etc. no change in the DNA occurs, even though a mutation did hit that location. These base pair mutations will not be counted as true silent base pair mutations). In a true silent point or base pair mutation one of the three nucleotides in the codon is exchanged for another in the mutation process. Even though this has occurred, the new codon codes for the same amino acid.

On the next page a careful examination of the tabulation of all 64 codons listed in Table 12-1 will show that silent point mutations usually occur when the third nucleotide (base pair) in a particular triplet is mutated to another one, which forms a different codon that codes for the same amino acid. There are several exceptions to this general rule. Two codons, namely those for methionine and tryptophan, have no other codons, which code for these amino acids. Therefore they cannot sustain a silent base pair mutation. One other exception involves the middle nucleotide for the stop codon ACT, which can be changed to another stop codon by a point mutation that exchanges the middle C for a T, making ATT, another stop codon. Two other exceptions involve changing the first nucleotide in TCT to G as in GCT and changing the T in TCC to a G as in GCC. All four of these codons code for arginine. Arginine has four other codons that code for it making a total of eight possible silent base pair mutations for arginine.

On the next page just above – Table 12-1 are displayed two separate examples of seven different DNA codons. In the first example, there are seven codons in a row, each of which is separated by a comma. To the right of this, a second example of DNA codons is displayed, exactly the same as the first, except that they have no commas in between each triplet codon. This is the way it really is in any DNA molecule, one triplet codon follows another with no separations in between. An analogy would be a sentence constructed with no spaces between each word or after a punctuation mark. It will be assumed that all of the codons noted in Table 12-1 are located in the anti-sense strand of a DNA molecule base and that complementary nucleotides, though not shown, will be present in the adjacent sense strand, thereby making base pairs.

AAA,AAC,AAT,AAG,GAA,GAG,GAT AAAAACAATAAGGAAGAGGAT

TABLE 12-1

A list of all 64 triplet codons is displayed below along with their respective amino acids for which they code. They are displayed in the same order as those in the boxed representations figure 3-2. However, each triplet codon is shown in the horizontal position instead of the vertical position as in figure 3-2.

Column Ns= # 0f non-silant mutations Column S= # of silent mutations

	Ns		S		Ns		S		Ns		S		Ns		S	
1	8	AAA	1	Phenyalanine	6	AGA	3	Serine	8	ATA	1	Tyrosine	8	ACA	1	Cysteine
2	8	AAG	1	________	6	AGG	3		8	ATG	1	________	8	ACG	1	________
3	8	AAT	1	Leucine	6	AGT	3		7	ATT	2	Stop	8	ACT	1	Stop .
4	8	AAC	1	________	6	AGC	3	________	8	ATC	1	Stop	9	ACC	0	Tryptophan
5	6	GAA	3	Leucine	6	GGA	3	Proline	8	GTA	1	Histidine	6	GCA	3	Arginine
6	6	GAG	3		6	GGG	3		8	GTG	1	________	6	GCG	3	
7	6	GAT	3		6	GGT	3		8	GTT	1	Glutamine	5	GCT	4	
8	6	GAC	3	________	6	GGC	3	________	8	GTC	1	________	5	GCC	4	________
9	7	TAA	2	Isoleucine	6	TGA	3	Threonine	8	TTA	1	Asparagine	8	TCA	1	Serine
10	7	TAG	2		6	TGG	3		8	TTG	1	________	8	TCG	1	________
11	7	TAT	2	________	6	TGT	3		8	TTT	1	Lysine	7	TCT	2	Arginine
12	9	TAC	0	Methionine	6	TGC	3	________	8	TTC	1	________	7	TCC	2	________
13	6	CAA	3	Valine	6	CGA	3	Alanine	8	CTA	1	Aspartic Acid	6	CCA	3	Glycine
14	6	CAG	3		6	CGG	3		8	CTG	1	________	6	CCG	3	
15	6	CAT	3		6	CGT	3		8	CTT	1	Glutamic Acid	6	CCT	3	
16	6	CAC	3	________	6	CGC	3	________	8	CTC	1	________	6	CCC	3	________
	110		34		96		48		127		17		109		35	= 576
	144				144				144				144			= 576 total

Total possible non-silent base pair or point mutations =110+96+127+109 = 442.

Total possible silent base pair or point mutations = 34+48+17+35 = 134.

From these calculations, 442/576= 0.767 or 76.7% and 134/576=0.233 or 23.3% it can be seen that 76.7% of all point mutations are non-silent and that 23.3% are silent.

Now that a silent base pair mutation has been defined above, a non-silent pair or point mutation is one that changes the code for a given codon in which it strikes. Table 12-1 lists the non-silent base pair mutations as Ns and the silent base pair mutations as S. Non-silent base pair mutations can be neutral, deleterious, or beneficial. A neutral base pair mutation changes a given codon to code for a different amino acid. However, the new amino acid acts so similar to the one that got replaced, that the resulting protein continues to function normally or nearly normal. A deleterious base pair mutation causes havoc in the resulting protein. A beneficial mutation is supposed to cause an improvement in the protein's function.

Because each codon in the transcribing strand is composed of three of the four kinds of nucleotides, it can be seen from the tabulations shown in Table 12-1 that there are 64 different kinds of codons, and each is composed of a different permutation of three of the four different kinds of nucleotides. This means that if 64 is multiplied by 3 a total of 192 individual nucleotides (base pairs) will be needed to compose all of the 64 codons in the genetic decoder. However, when non-silent and silent point mutations are added together, there will be a total of 576 different point mutations possible. This is because for any one of the three nucleotides in any DNA codon there exist three possible base pair mutations. In other words, A can be mutated only to T, G, or C. Therefore A cannot be mutated to itself. The same applies to each of the other three base pairs making a total of 9 possible base pair mutations for each codon. (Actually as previously noted, any one of the four bases could sustain a mutation of itself, which would produce no change. However, they are not counted as silent base pair mutations). When all the Ns columns are totaled, the sum 442 (or 76.7%) is the number of possible non-silent base pair mutations in all 64 codons. The summation of all the S columns 134 (or 23.3%) denotes the total number of possible silent base pair mutations in all 64 codons. Ns mutations occur 3.39 times more often than S mutations.

This attempt to discover how a prokaryote could evolve into a single-celled eukaryote will follow only one of many possible evolutionary pathways. Prokaryotes, unlike eukaryotes, have very few if any introns or non-coding segments in their DNA. Because of this it becomes apparent for practical purposes that each succeeding triplet codon in the circular DNA molecule in a prokaryote has to be attached to a previous codon with no large non-coding sections (introns) in between. This means that for practical purposes, most random point mutation will strike a triplet, which codes for a specific amino acid with each occurrence. In addition it must be noted that any random point mutation occurring anywhere in any DNA molecule in a prokaryote can only hit on the first, second, or third nucleotide of the codon that it strikes. Therefore, when any point mutation randomly strikes any codon, there will be a 33 1/3 percent chance that any one of the three nucleotides composing any codon will be hit by any given point mutation.

Remember from a previous chapter that the correct composition of any protein depends upon the specific sequence of the various combinations of the 20 biological amino acids. The number of amino acids in a given protein can vary from about 100 on up to greater than 1000. Not all proteins contain all 20 biological amino acids, just like not all English sentences contain all 26 letters of the alphabet. However, just like all correctly spelled words in an English sentence must be in a specific sequence in order to make sense, so also correctly constructed proteins must have the correct sequence of biological amino acids, whether or not all 20 amino acids are needed in a given protein. The correct sequence for each amino acid in every biological protein is derived from the correct sequence of each codon in the DNA of a given gene.

Many single non-silent base pair mutations are deleterious, causing such havoc in the gene where they happen, that they produce a lethal result in the individual organism in which they occur. They are known as missense base pair mutations. However, some non-silent base pair mutations occurring in specific places in genes that code for certain proteins produce very little if any impairment to the function of that particular protein, these are called neutral mutations.

In fact, it has been estimated that some proteins can have up to about 66% of their amino acids changed with very little change noted in their function. This phenomenon is noted in the cytochrome C proteins which are ubiquitous to all living organisms from bacteria to humans. Cytochrome C molecules are essential to all living organisms because they are involved with oxidation processes used by all biota to obtain energy to run their respective metabolisms. Most of these cytochrome C proteins contain approximately 100 amino acids, which would have been translated from about 100 codons composed of 300 base pairs in the DNA. It has been estimated that only about one third of the 100 amino acids are absolutely essential both as to kind and placement in this molecule for function to exist. Any non-neutral base pair mutation that occurred in this one-third (about 100 base pairs) of this molecule would be deleterious and fatal to the organism in which it took place. On the other hand, this means that approximately two-thirds of the remaining base pairs (about 200) can vary considerably with virtually no change in function of this very essential protein. An exception of course would be the accidental production of a stop codon. Because approximately two-thirds of the amino acids (both as to kind and placement) used in construction of this protein are not absolutely essential to the function of this very important protein for life, it should come as no surprise that when this molecule is analyzed from various species of biota, that great differences both as to kind and placement of many of the amino acid will vary. This information was published in 1972 by M.0.Dayhoff in the Atlas of Protein Sequences and Structure, National Biomedical Research Foundation, Silver Springs, Maryland, volume 5, Matrix 1, p D-8.

The cytochrome C molecule described above is an exception to the rule. Many proteins cannot tolerate a 66% exchange in their amino acid structure without losing their function. Some proteins lose their function with only one or two base pair mutation changes, especially if the mutation strikes in a critical area.

The goal here is to try to proceed along one reasonable possible pathway that evolution could have taken while evolving a single-celled eukaryote from a hypothetical prokaryote. The minimum prerequisites for biological evolution to occur are the presence of DNA, RNA, and ribosomes in the metabolism of any organism. It is in these ultra microscopic structures of DNA or RNA where mutations accidentally happen. Mutations change the genetic instructions that the ribosome then translates. According to Kimball's Biology Pages (www.biology-pages.info), the E. coli bacterium contains 4,639,221 base pairs with 4377 genes. These are all in one chromosome. For mathematical simplicity, a hypothetical primordial prokaryote will be considered in the remaining portion of this chapter. It will be assumed to have 4.5×10^6 base pairs and 4500 genes. Its complete replication cycle will include a lag phase, an exponential replication phase, a stationary phase, and a death phase. It will be able to replicate itself every 30 minutes when in the exponential replication phase and on average will sustain one base pair mutation with every 10^9 base pair replications. During the death phase the bacterial population will decrease exponentially at the same rate as it increased during the exponential replication phase. As noted in Table 12-1, a base pair mutation can only be one of three different kinds. Therefore, each of the 4.5×10^6 base pairs can randomly be struck by any one of the three different kinds of base pair mutations. This means that there are $3x4.5x10^6$ or $1.35x10^7$ different base pair mutation possibilities that can occur in this hypothetical organism.

Because prokaryotes are known to have very few if any introns, we can obtain the approximate average number of base pairs in a gene of the hypothetical primordial bacterium by dividing $4.5x10^6$ by 4500 which yields 1000. To make the math easier, we will round off the 1000 base pairs in an average gene to 999, this way the number of base pairs will be devisable by 3. This means that there would be 333 codons in the average prokaryotic gene.

It is obvious that because evolutionary processes can have no goals, any advancement in biological complexity if reached by evolution, would of necessity be an accident. If evolutionary processes produced the first single-celled eukaryote, it would have been a distant accidental descendent of the first living organism, most likely a prokaryote. With no pre-designated goals, this evolutionary feat is supposed to happen without the aid of memory or intelligence from random mutation accidents accepted or rejected by natural selection.

If evolution is true, there is no way to know what the first single-celled eukaryote would have been. However, a fungus (yeast) by the name of Saccharomyces cerevisiae, one of the simplest single-celled eukaryotes known to exist, would be a good candidate to represent one of the first to have evolved. Again according to Kimball's Pages sighted above, it contains 16 paired homologous chromosomes in a total of 12,495,682 base pairs coding for 5770 genes. For mathematical ease, the number of base pairs will be rounded to $1.25x10^7$ and the genes to 6000. Therefore, this hypothetical single-celled eukaryote will have 15 more chromosomes, about 8 million more base pairs and about 1500 more genes than the hypothetical primordial prokaryote has in its genome. It will be assumed that the average number of base pairs in the average gene in this simple hypothetical yeast, will also contain approximately 999 base pairs, although obviously this number could be more or less. It is known that most eukaryotes have non-coding DNA interspersed between their genes.

Accordingly, evolutionary processes will have 2 billion years to correctly mutate the base pairs, for modification of the 4500 original genes needing remodeling. During this same time, these processes will have to build the 1500 additional new genes. To understand just how slow evolutionary processes are, if 8 million, the number of additional base pairs in a single celled-eukaryote as compared to the original prokaryote is divided into 2 billion, we get 250. This means that on the average one new base pair would be added about every 250 years to the genome of the original hypothetical primordial prokaryote. Using similar calculations, each one of the individual 1500 new genes would on average make its appearance about every 1,333,333 years. Sometimes a given evolutionary event might occur in less time but on the average, according to this scenario, evolution would proceed very slowly, just like Darwin predicted.

At the outset, in our attempt to understand evolutionary advancement, mutations associated with meiosis and sexual reproduction would not be present in the primordial prokaryote because prokaryotes reproduce asexually. However, theoretically after the eukaryotes appeared on the biological scene, sexual reproduction appeared in a simple form as noted in some single-celled eukaryotes. Therefore beneficial mutations that might occur with meiosis would not be available for evolution's theoretical advancement of a hypothetical prokaryote to a single-celled eukaryote.

Because evolutionary theory postulates that the prokaryote was the progenitor of the single-celled eukaryote, through a long line of intermediates, it must be assumed that the 4500 genes originating in the hypothetical prokaryote eventually became part of the genome of the hypothetical single-cell eukaryote. Of course this would mean that most, if not all, of these 4500 genes had to be modified in various ways in order that they improve function in the evolving prokaryote and later be useful in the single-celled eukaryote as well. Arbitrarily, it will be estimated in this evolutionary process that on the average 15% of the base pairs of each of these 4500 genes will need modification, and if we say that the average gene in the prokaryote contained 999 base pairs, then this would mean that approximately 150 base pairs on average would need to be exchanged for others in each gene during this evolutionary modification. The base pairs involved in these modifications need not be all contiguous in a given gene although some or all of them could be. This 15% estimation is probably very conservative. The percentage most likely should be greater. Fifteen percent of 4.5×10^6 is 675,000. This is a conservative estimate of the total number of base pairs that must be randomly and correctly mutated in this hypothetical prokaryote in order for it to evolve toward a hypothetical single-celled eukaryote.

Also, during this process, the single circular chromosome of the prokaryote must be broken up into 16 haploid chromosomes and 16 homologous chromosomes must form and become associated with the broken ones. Also during this time, the evolutionary processes must manufacture the organelles and the size of the organism itself must increase tremendously, in some cases up to 10,000 times in volume.

Consequently, for the random evolutionary advancement of a single-celled eukaryote from a prokaryote, the most difficult problems are; #1, the rearrangement of existing base pairs that will have to occur in the genome of the prokaryote; #2, the addition of millions of base pairs and their use in the formation of about 1500 new genes; #3 the splitting of the single circular chromosomes of the prokaryote into multiple fragments to form the 16 individual noncircular chromosomes found in the hypothetical single-celled eukaryotes; #4, the construction of homologous chromosomes to form pairs with each of the 16 fragments; #5, the origin of organelles present in all eukaryotes but not found in prokaryotes; and #6, the marked increase in size of eukaryotes over the size of prokaryotes. The mathematical calculations which follow, will involve #1 only.

So now we need to calculate the time it would take and what the chances would be for random base pair mutations to correctly modify the 4500 original prokaryotic genes. In addition, no attempt will be made to mathematically solve the problem regarding the formation of 1500 new genes, each added one or two at a time to the genome of a prokaryote as it evolves toward a single-celled eukaryote. However, as the organism evolves, it must of necessity pass through thousands or possibly millions of intermediate life forms. In addition, all of this must occur in a two billion-year timeframe by accidental random means with no intelligence, memory, or goal.

The following "Givens" form the basis for one way this evolutionary accident might have occurred.

Given: 1. At the outset, the hypothetical primordial prokaryote will contain 4.5×10^6 base pairs with 4500 genes contained therein.

Given: 2. The number of base pairs per gene as previously estimated in this chapter will on average be 999 with 333 codons per gene.

Given: 3. This hypothetical primordial prokaryote will be able to replicate itself every 30 minutes under optimal conditions during the exponential replication phase of its replication cycle.

Given: 4. This hypothetical primordial prokaryote will sustain on the average one base pair mutation per 10^9 base pair replications.

Given: 5. This primordial prokaryote will replicate by simple cell division, which includes a lag phase, followed by an exponential replication phase, followed by a stationary phase, and lastly followed by a death phase. During the lag phase and the stationary phase bacterial replication will be very slow. During the exponential replication phase, bacterial population growth will follow the equation $P=2^n$, where P equals population census at any given replication n. During the death phase the bacterial population will decrease exponentially at the same rate at which it was built up (See table 12-2).

Given: 6. The hypothetical single-celled eukaryote (a yeast) will have about 1.25×10^7 base pairs in its genome, which will contain about 6000 genes. This will be considered as an artificial accidental goal for the evolution of the hypothetical prokaryote to a hypothetical single-celled eukaryote.

Given: 7. Each of the 4500 existing genes in the prokaryote will also require on average a 15% remodeling of their base pairs. Fifteen percent of 999 equals about 150 base pair changes per gene.

Given: 8. As it evolves toward the hypothetical single-celled eukaryote, evolutionary processes must create and add 1500 new genes to the genome of the hypothetical prokaryote. However, no attempt will be made to calculate the time required for the formation of the additional genes.

Given: 9. Silent mutations compose 23.3% of the base pair mutations, leaving 76.7% as a non-silent.

Given:10. According to the estimates of paleontologists, it took evolution about two billion years to evolve a prokaryote into a single-celled eukaryote.

The following mathematical calculations, will use the above factors in an attempt to determine the probability for evolutionary processes to modify the 4500 original genes of the prokaryote into those usable for a single-celled eukaryote. Also an attempt will be made to determine the time it would take for this to occur. The first problem involves the probability of the modification by random base pair mutations of the 4500 original genes of the prokaryote. The second problem will be an attempt to find out how long it would take. Each gene must be modified in such a way as to make it beneficial not only to the prokaryote in which it is being modified but also beneficial to a single-celled eukaryote that may inherit it many years later.

There are other factors that cannot be entered into the calculations, which follow but both must be considered. Each has to do with the problems of succession. After the first of the 150 individual correct base pair mutation sites is located in one of the 4500 original genes, it becomes progressively harder and harder to find each successive correct mutation site in that gene. This is because the ratio of finding the first correct mutation site that will need to be correctly mutated, will be one out of 150 located somewhere in $4.5x10^6$ base pairs. However, the next correct mutation will be one out of 149 located in $4.5x10^6$ base pairs, then one out of 148 and so on down till the last one will be one out of $4.5x10^6$ base pairs. In other words, the degrees of freedom decrease with each successful mutation. In each of the 4500 original genes, it will take a longer time for random mutations to locate each successive correct mutation site as the number of possible correct mutation locations decreases. This factor could be entered but would only make the calculations more complicated and it would increase the time needed to modify each of the 4500 genes. Each incorrect mutation site must be rejected by natural selection. No attempt will be made to take this into consideration.

The second problem in the calculations which follow has to do with the fact that no significance is given to the order in which the mutations occur. Therefore, any one of the 150 base pairs that need modification in each of the 4500 genes could be modified in any order just so long as any one of the 150 correct mutation sites is correctly mutated. In the calculations, no priority is given as to which one of the 150 base pairs will be selected first nor any order of selection of any one of the remaining 149 base pairs there-after. When building organic molecules, priority must be given as to which portions in succession must be added to the molecule. The order in which portions of the molecule are joined to each other is very important. This would be analogous to building a house. First, the forms for the foundation and cement slab must be made. Then the plumbing pipes and electrical conduits must be placed. Next comes the pouring of the concrete for the foundation and floor, followed by the framing, followed by the roof, and last the siding, etc.

This same idea applies to building an organic molecule. Priority must be given to the order in which additions are made. The roof of a house cannot be made first and then lay the foundation, followed by building a house in-between. The order of succession of each base pair exchange needed in modifying each gene in the 4500 original ones will be different and unknowable. Therefore, each base pair mutation in each gene must occur in a different specific order. Some genes may need huge amounts of modification and others may need very little. Because not only the order but also the amount of modification, which each individual gene must undergo is unknowable, these factors cannot be entered into the calculations which follow. However, it is easy to see that if it were possible to add them, they would significantly decrease the probability of their occurrence and increase the time that it would take to happen.

By definition, we know that evolutionary processes are very slow, just as Darwin insisted that they should be. Each step toward more biological complexity, according to him, should be the shortest that can be taken. Attempts will be made to mathematically demonstrate just how slow evolution really is. For the readers who are not mathematically inclined, they may skip forward to the conclusions shown in bold type several pages ahead.

THE ASSUMPTIONS USED FOR MODIFYING THE 4500 EXISTING GENES

1. *A* hypothetical simple one-celled organism (a prokaryote) replicates exponentially—its population doubling every 30 minutes under favorable environmental conditions. Under these conditions, starting with one organism, the population P after n generations (n/2 hours) is given by the equation $P = 2^n$.
2. Each such organism has $4.5x10^6$ base pairs each of which, in each replication of the organism, replicates to itself or mutates to one of three other base pairs. This results in $3x4.5x10^6=1.35x10^7$ potential mutation sites for the organism in each replication.
3. Each of the $1.35x10^7$ potential base pair mutation sites is equally likely to occur.
4. Of the $1.35x10^7$potential base pair mutations, 23.3% are silent, that is, they have neither a detrimental nor a beneficial effect on the organism. The remaining 76.7% are non-silent and except for a few neutral mutations are detrimental or beneficial in their effect on the organism.
5. On average, one base pair mutation occurs in every 10^9 base pair replications.
6. Mutations are mutually independent, that is, the occurrence or non-occurrence of a mutation at any base pair does not influence the occurrence or non-occurrence of a mutation at any other base pair.

Think of these $4.5x10^6$ base pairs lined up in a long line. Arbitrarily select any 150 of them. Upon the basis of the above six assumptions, we want to calculate the probability that a specific mutation occurs in a particular one of the 150 base pairs with the 149 remaining mutations occurring in any order, and no mutation occurs at any of the remaining $4.5x10^6$-150 base pairs in the organism.

THE PROBABILITY OF THE FIRST CORRECT BASE PAIR MUTATION OCCURRING

P(a particular non-silent mutation occurs at one of 150 selected base pairs in a given gene in an organism having $4.5x10^6$ base pairs and no mutation occurs elsewhere in the organism)
=P(mutation occurs)xP(it is non-silent)xP(it occurs at one of the 150 selected base pairs)
xP(it is a particular one of the 3 possible mutations at each base pair)
xP(no mutation occurs at any of the remaining $4.5x10^6$-1 base pairs in the organism)
$= (1/10^9)x0.767x(150/(4.5x10^6))x(1/3)x((10^9-1)/10^9)^{(4.5 x 10\wedge 6 -1)} = 1/(1.1787x10^{14})$
We would thus expect a specific mutation to occur at one of the 150 selected base pairs with no mutation occurring at any of the remaining $4.5x10^6$-1 base pairs in the organism, on average, in

every $1.1787x10^{14}$ base pair replications. Since the organism has $4.5x10^{6}$ base pairs, this should happen on average, once in every $1.1787x10^{14}/4.5x10^{6}=2.619x10^{7}$ organisms.

AVERAGE NUMBER OF GENERATIONS FOR THE 1ST MUTATION TO OCCUR

We now want to calculate how many organism generations it takes to obtain $2.619x10^{7}$ organism replications. In the first organism generation there would be $1=2^{0}$ organism replications. In the second generation there would be $2=2^{1}$ organism replications; in the third $4=2^{2}$; in the forth $8=2^{3}$;... ad infinitum. We want the sum of all these organism replications to equal $2.619x10^{7}$. Adding these generational organism replications we get the following: $2.619x10^{7} = 2^{0} + 2^{1} +2^{2} +2^{3} +\ldots+ 2^{n-1}$

But the sum of this geometric series is 2^{n}-1, so we must solve $2.619x10^{7} = 2^{n}-1$. Doing this we get $2^{n} = 2.619x10^{7}+1$ -> n x In2 = In($2.619x10^{7}$+1) -» n = In($2.619x10^{7}$+1) In2 = 24.64 (See table 12-2). This means it will take, on average, 24.64 exponential generations of the organism to obtain $2.619x10^{7}$ organisms, which results in $2.619x10^{7}$x $4.5x10^{6}$ = $1.1787x10^{14}$ base pair replications. This in turn means we would expect, on average, the first correct (beneficial) base pair mutation to occur at one of the selected 150 base pairs in one specific gene in one organism with no mutation occurring at any of the remaining $4.5x10^{6}$-1 base pairs in that organism. On average, this should occur once in the first 24.64 generations of the original organism. Assuming the organism population doubles every 30 minutes during the exponential replication phase. 24.64 generations would take, on average, 24.64/2 = 12.32 hours.

Even though only one correct (beneficial) base pair mutation occurred in only one specific bacterium in the first 24.64 generations, which produced a population of $2.619x10^{7}$ bacteria, there were actually 90,405 additional mutations that occurred in other individual bacteria at the same time. A population of $2.619x10^{7}$ bacteria when multiplied by $4.5x10^{6}$, the number of base pairs in one bacterium, will yield a total of $1.1787x10^{14}$ base pairs. Remember, on the average, one base pair mutation occurs with every 10^{9} base pair replications. Therefore, the total number of base pair mutations can be calculated by dividing $1.1787x10^{14}$ by 10^{9} which yields $1.1787x10^{5}$. This number must be multiplied by 0.767, which gives the fraction of non-silent base pair mutations. This yields 90,406, which is the total number of bacteria containing non-silent base pair mutations that occurred in the first 24.64 generations. However, only one of the 90,406 non-silent base pair mutations that occurred during the exponential replication of the $2.619x10^{7}$ bacteria is correct. Aside from neutral base pair mutations, which compose part of this number, for the balance of, the 90,405 other base pair mutations would most likely be deleterious and therefore eliminated by natural selection.

It must be pointed out again at this juncture that the prokaryotic replication cycle contains the lag phase, followed by the exponential replication phase, which in turn is followed by the stationary phase, and finally the death phase. During the death phase the bacterial population decreases exponentially at the same rate at which the population increased during the exponential replicating phase of the replication cycle. Obviously, some bacteria escape almost every death phase, or else all bacteria would then soon die off. So during a given death phase, the chance for the one correctly mutated bacterium to escape into the environment and not die, would be one in 26,193,333. Therefore, it would take 26,193,333 complete replication cycles (which includes all four phases mentioned above) to be quite sure from a probability point of view, that one correctly mutated bacterium would be able to escape into the environment. Because the correctly mutated bacterium could have escaped into the environment during the death phase of the first replication cycle or the last, 26,193,333 must be divided by two, which yields 13,096,666. Without figuring the standard deviations into the calculations, this means that on the average, during many death phases, the correctly mutated bacterium would have one chance in 13,096,666 complete replication cycles of escaping into the environment. If it did escape it would then go through a lag phase before it could find another friendly environment in which it could start replicating exponentially again. Therefore, the most likely kind of bacterium to escape into the environment would be one that is not mutated. The bacteria with the deleterious mutations would most likely have already been eliminated by natural selection. However, on average one correctly mutated bacterium should escape into the environment during one of the 13,096,666 completed replication cycles.

During this attempt to calculate the time needed to evolve this hypothetical bacterium towards a single-celled eukaryote, the exponential replication process was artificially stopped when the bacterial population reached 26,293,333. This had taken12.32 hours. Following this artificial stopping point, the stationary phase would begin. Its average length in time is not known. The death phase would then follow, which would take about the same length of time as the exponential replication phase, namely about 12.32 hours. Then comes the lag phase, its time varies between minutes on up to days or possibly even longer. The stationary phase would probably be about the same length regardless of how many exponential replication cycles occurred. The same could not apply to the lag phase, because when any bacterium escapes the death phase and moves into the environment, it may find another friendly place to replicate rather quickly, or it might take days. Arbitrarily, the length of the lag phase will be set at 12 hours. This is very conservative, probably being way too short. Therefore it is very unlikely that any complete replication cycle will follow another so quickly. Arbitrarily, the stationary phase will be set at 3.36 hours in length. When added together, 12.32+12.32+3.36+12 makes a total replication time of 40 hours.

Bacteria sometimes can exponentially replicate longer than 12.32 hours. Therefore, this would increase the length of a given replication phase. Obviously, this would depend upon the increased number of exponential replications. The increase in length of a given exponential replication phase will also increase the length of the death phase by approximately the same amount.

After the colony reaches 24.64 exponential generations, the number of so-called beneficial mutations will also double with each additional replication and so will the non-silent mutations. But remember that no matter how many complete replication cycles take place, the probability ratio for any correctly mutated bacterium to escape a given death phase will remain constant at one in 13,096,666 replication cycles. This one correctly mutated bacterium can be one of many correct ones but regardless, because as mentioned above, every time the bacterial population increases by 26,193,333 there will also be approximately a total of 90,406 base pair mutations, only one of which on average will be correct. So if a given bacterial colony finds itself in a very friendly environment and continues to replicate exponentially more than 24.64 times following the equation $P=2^n$, every time the population increases by 26,193,333, the probability will be that one correct base pair mutation will have occurred along with 90,405 others. Of course, the probability for one correct non-neutral base pair mutation to occur would be much less if the exponential replication phase did not reach 24.64 replications, which frequently it doesn't.

If the 24.64^{th} replication is chosen as an endpoint, the replication cycle according to this estimate will require 40 hours to complete. This endpoint represents the shortest number of replications in which we could expect at least one correct non-neutral base pair mutation to occur. Any number of replications longer than 24.64 replications would produce more replication cycles and with more correct mutations, but the ratio of correct mutations to the number of base pairs replicated would remain constant. In addition, the ratio of bacteria containing the correct mutation, which escape the death phase, will also remain constant when compared to those that do not have the correct mutation. So how long in years, would it take for the random processes described above, to produce on average 150 correct or beneficial base pair mutations in 4500 genes? To find out, the above factors must be multiplied together and then divided by 8766, which equals the number of hours in one year. Multiply $13{,}096{,}666 \times 150 \times 4500 \times 40 = 3.5361 \times 10^{14}$ hours. **Divide this last number by 8766 and the answer is 40,338,807,000 years. This shows that evolutionary processes dependent on random means, with no intelligence, memory, or goal, could not reach this accidental goal in 2 billion years using hypothetical numbers, which in most cases were derived as close as possible to real ones and for the most part are quite conservative.**

If the theory of evolution is true, evolutionary modifications at least similar to those shown above, had to have occurred in the 4500 genes original to the prokaryote. By isolating this one aspect of evolutionary change, a rough mathematical analysis of it has been obtained. Obviously, during the same time that the gene modifications were theoretically in progress, additional changes had to be ongoing. These included the addition of eight million more base pairs to the genome. Some of them were used to form the 1500 new genes, some for the doubling of the originial 4500 genes during making of the homologues chromosomes, and some for the introns. Simultaneously, the single circular chromosome of the prokaryote had to be divided into 16 linear chromosomes, along with formation of 16 homologous chromosomes making the16 homologous pairs. The formation of the homologous chromosomes used up a lot of the 8 million additional base pairs. All of these changes required the original prokaryote cell size to increase markedly. No attempt was made to mathematically analyze how long it would take for any of these multiple other random evolutionary events to occur including origin of the organelles.

The modification of the 4500 genes did not significantly increase the number of base pairs in the genome in the prokaryote. Basically, this modification process only exchanged thousands of base pairs for others in the genome of the prokaryote. Beside the factors used in the calculations, don't forget that two others dealing with the order of succession were not used. They were described earlier in this chapter. If these would have or could have been utilized in the calculations, the time needed for the 4500 gene changes, would have exceeded far beyond 40 billion years. The time that evolution would require for the gene modifications alone, shows how improbable it would be for random evolutionary processes without memory, intelligence, or goal to have produced these changes in the time allotted. **From these approximate calculations, it can be seen that two billion years is not long enough for random evolutionary processes to produce the changes necessary in that length of time.**

So what can we learn from this mathematical analysis? It addresses only one small fraction of all the changes that would be required to evolve a single-celled eukaryote from a prokaryote. Darwin had imagined that evolutionary complexity would advance using the smallest possible steps. Single base pair mutations are the shortest steps and they were used in calculating the modifications of the 4500 genes. It becomes apparent that mutations larger than base pair mutations would have a greater chance of introducing havoc into a given gene, than changing the gene slowly with single base pair mutations.

In addition, the calculations seem to show that besides not having enough time, one additional boundary impeding accidental evolutionary progress was the inability of the exponential replication process of the prokaryote to sustain itself for much more than about 25 hours to 50 replications, because the weight of the colony starts to become too large. See Table 12-2. It seems probable that the chances for increased evolutionary complexity, produced solely by random means, would have a better chance if the exponential phase of bacterial replication could be extended for long periods of time.

However, this is not true either. The lack of any significant evolutionary increase in complexity is demonstrated by the continuous exponential replication of the common prokaryote E. coli. Among the many places where this bacterium seems to thrive, is a unique ecological niche, the distal small bowel of man and other warm-blooded animals. This warm liquid environment provides an ideal medium in which E. coli can continuously replicate exponentially. As fast as the partially digested food moves downstream in the gut, the bacteria replicate exponentially and swim upstream. This is a continuous dynamic process. In any other ecological niche, if E. coli could continuously replicate exponentially, they would take over the entire world in less than 72 hours. The way this theoretically could happen was demonstrated earlier in this chapter. However, in the terminal ileum, as fast as they proliferate they pass out in the feces. Here, they soon go into the death phase followed by the lag phase. Depending upon where the feces land, the lag phase may last for months. In developed countries such as the United States, human feces containing E. coli end up in the sewer system, where theoretically they remain trapped.

A quotation from the Encyclopedia of Molecular Biology (Kendrew, page 337, published 1994) seems to emphasize this point. "E. coli are present in the gut within a few days of birth and throughout life. Thus even if only the population of E. coli in man is considered the number of individual cells present in the world at any one time, though difficult to estimate accurately, must be very large. The mutation rate, ~1bp change per 10^{10} bp replicated (i.e. bp per 200 replications), would suggest that there ought to be a great many variants but this is not found. Most of the isolates obtained in a short period of time can be grouped into relatively few major clones. This implies that the pressures which maintain the homogeneity of the species are highly selective." E. coli living in the terminal small bowel must replicate at an exponential rate continuously in order to maintain their presence at the head of the fecal stream. As fast as these bacteria replicate, they are being swept out in the feces.

The importance of this concept, from an evolutionary point of view, is that even though these bacteria are continuously replicating at the exponential rate, very little if any significant increase in evolutionary complexity occurs. When one considers the huge number of these bacteria present in the world at any one time living in the guts of humans and warm-blooded animal's, the number is staggering. Also keep in mind that this exponential replication has been going on continuously, 24 hours a day, seven days a week for millenniums. There should be an astronomical number of theoretical beneficial mutations being produced every day in this group. Supposedly, they should lead to more complexity. Yet as noted above, significant variants are absent. The billions of ecological niches, resident in warm-blooded animals and humans worldwide, should be an ideal place for random evolutionary processes to produce in E. coli accidental random changes toward more complexity but they haven't.

Consideration must now be given to the problems that would arise during evolutionary addition of the 1500 new genes to the genome of a prokaryote, while it is theoretically advancing toward a single-celled eukaryote by accident. We have learned from transpondable element mutations, which occur when a gene in one part of a chromosome is moved to another place that the transported gene in its new location influences the expression of genes in that vicinity. Therefore, we can infer from this that there has to be a specific order of succession in which each of the 1500 new genes can be added to the DNA so as to prevent havoc in gene expression.

Additional genes, homologous chromosomes, and introns all require additional base pairs to be added to the genome of the prokaryote, as it is evolving toward the single-celled eukaryote. The transformation process discussed earlier this chapter, occurs in prokaryotes. It depends upon naked DNA being released in the environment due to the death phase of a bacterial colony, with lyses of individual bacterial cells. Sometimes this naked DNA is absorbed by a live prokaryote and becomes incorporated into its genome. This seems to be the easiest and most accessible method for base pairs to be added to a given prokaryote's genome. Once additional base pairs have been added in increments to the genome of an evolving prokaryote, the likelihood that they could be modified by random base pair mutations into one or more new genes over time seems possible.

The mathematical improbabilities developed in this chapter, may possibly be overcome by some future scientific discovery, but a brake-through of this magnitude seems very unlikely.

Table 12-2

1. When n= 1, P=2
2. When n= 2, P=4
3. When n= 3, y=8
4. When n= 4, P=16
5. When n= 5, P=32
6. When n= 6, P=64
7. When n= 7, P=128
8. When n= 8, P=256
9. When n= 9, P=512
10. When n=10, P=1,024
11. When n=11, P=2,048
12. When n=12, P=4,096
13. When n=13, P=8,192
14. When n=14, P=16,384
15. When n=15, P=32,768
16. When n=16, P=65,536
17. When n=17, P=131,072
18. When n=18, P=262,144
19. When n=19, P=524,288
20. When n=20, P=1,048,576
21. When n=21, P=2,097,152
22. When n=22, P=4,194,304
23. When n=23, P=8,388,608
24. When n=24, P=16,777,216
25. When n=25, P=33,554,432
26. When n=26, P=67,108,864
27. When n=27, P=134,217,728
28. When n=28, P=268,435,456
29. When n=29, P=536,870,912
30. When n=30, P=1,073,741,824
31. When n=31, P=2,147,483,648
32. When n=32, P=4,294,967,296
33. When n=33, P=8,589,934,592
34. When n=34, P=17,179,869,180
35. When n=35, P=34,359,738,370
36. When n=36, P=68,719,476,740
37. When n=37, P=137,438,953,500
38. When n=38, P=274,877,907,000
39. When n=39, P=$5.497558129 \times 10^{11}$
40. When n=40, P=$1.099511625 \times 10^{12}$

50. When n=50, P=$1.125899909 \times 10^{15}$

60. When n=60, P=1.1529215×10^{18}

This last number of bacteria resulting from only 60 exponential replications occurring at the rate of 2 per hour for 30 hours would weigh 115,292 gms. This would be a huge colony, almost too big to imagine.

The Second Overhanging Cliff
Bacteria to Single-celled Eukaryotes

It is extremely improbable for evolution to randomly evolve a single-celled eukaryote from a prokaryote in the time available.

Listed below are five reasons why blind evolution, without memory, intelligence, or goal could not make the second leap from prokaryote up to single-celled eukaryote.

1. There are millions of additional base pairs in single-celled eukaryotes as compared with prokaryotes. This only protrays the difference in total numbers of base pairs but says nothing about the thousands of additional genes needed. Neither does this address the thousands of rearrangements needed to change prokaryote genes into eukaryote genes.

2. The time available (2 billion years) is not long enough from a mathematical point of view to randomly produce the millions of changes needed to evolve single-celled eukaryotes from prokaryotes.

3. Evolutionary theory has no real explanation of how the single circular chromosome of the prokaryote is changed into multiple paired chromosomes of the single-celled eukaryote.

4. Evolutionary theory has no reasonable explanation for the origin of the multiple different kinds of organelles found in the single-celled eukaryote and not present in prokaryotes.

5. All single-celled eukaryotes are much larger in volume than the largest prokaryote, even up to 10,000 times larger. This larger volume requires transport systems to move products of metabolism from one site in the cell to another. Tiny molecular motors carry this out in eukaryotes in contrast to diffusion and concentration gradients which do the job in prokaryotes. Evolutionary theory has no explanation for the origin of these motors.

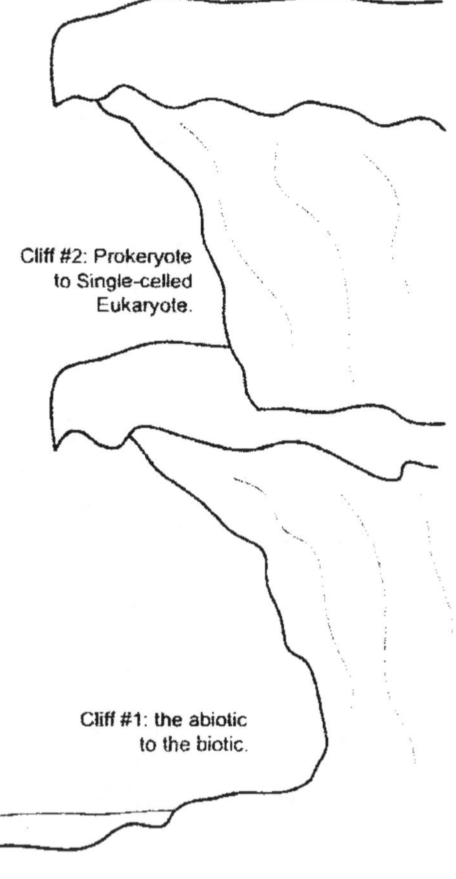

Figure 12-2

The genetic distance between the prokaryote and the single-celled eukaryote is gigantic! The genetic distance between prokaryotes and eukaryotes can be likened to two islands of organization separated by a huge sea of chaos. Imagine the number of molecules of water separating one of the Aleutian Islands in the North Pacific Ocean from New Zealand thousands of miles to the south. Then think of each island as the genetic distance that exists between prokaryotes and single-celled eukaryotes. This is not an exaggeration, because without intelligence, memory, or goal to facilitate these accidental theoretical evolutionary advances, an actual infinity of deleterious mutations could occur repeatedly, which natural selection would repeatedly have to reject an infinite number of times without any progress being made. Even if the selection pressures changed during the tenure of a given beneficial mutation, there would still be an infinite number of possible random deleterious mutations, any one of which might produce no progress at all or could destroy the benefits of an existing beneficial mutation.

With all of the improbabilities noted above, it is obvious that evolutionary processes could not overcome them by random means which are blind, have no intelligence, no memory, and no goal. Evolution cannot climb the overhanging cliffs (Figure 12-2) even with repeated tries and 2 billion years to do the job. As far as permutations of base pairs and/or codons in DNA are concerned, there is almost an infinite order of magnitude more combinations of disorder then there are combinations of order. As suggested by Stephen Hawking, this problem can easily be visualized by repeatedly shaking up pieces of a jigsaw puzzle in a box, and repeatedly expecting to see the picture resulting from perfectly placed pieces of the puzzle altogether each time the box is opened. But alas, with each opening, a different combination of disorder is noted even though there is the exact number of pieces with each piece designated to fit into another piece perfectly. This is a similar analogy to advancement proposed by evolutionary theory, except that the evolutionary scenario is many orders of magnitude less likely to occur. This is because there are many orders of magnitude more pieces in the genetic puzzle that must come together in perfect sequence by pure chance than there are pieces in the jigsaw puzzle. Laws of probability do not prohibit all molecules of gas in a balloon to simultaneously be all located in one-half of the balloon, allowing the other half to go flat. However, the chance for this to occur for practical purposes is as close to zero as you can get. This same notion applies to evolutionary theory.

SUMMARY

1. Single-celled eukaryotes are up to ten thousand times larger in volume than prokaryotes.
2. Eukaryotes have multiple kinds of organelles which, to the eukaryotic cell, would be analogous to organs of multi-celled eukaryotes such as heart, kidney, liver, etc. Prokaryotes have no organelles.
3. Except for the circular chromosome located inside the two organelles, the mitochondrion and the chloroplast, the eukaryotes have paired homologous chromosomes located in the nucleus. This is in contrast to the single, circular chromosome of the prokaryote, which is not located inside of any nucleus.

4. In eukaryotes the cytoplasm surrounding the nucleus is divided by various partitions each providing various functions.
5. Internal skeletal structures provide support for the eukaryote as contrasted to the tough cell wall of the prokaryote.
6. There are transport systems the eukaryote carries out with tiny "molecular motors" to move new parts manufactured and waste products to be excreted. The prokaryote depends upon diffusion and concentration gradients to perform these tasks.
7. Information just mentioned above was not available to Darwin in 1859.
8. E. coli, a common prokaryote, contains about 4.5 million base pairs in its genome.
9. Saccharomyees Cerevisiae, a common single-celled eukaryote, contains more than 7 million more base pairs in its genome than E. coli.
10. The trial and error method of evolution supposedly is able to evolve a prokaryote into a eukaryote by a method that has no memory, intelligence, or goal.
11. Evolutionary theory proposes that mutations accepted or rejected by the ever-present judge-- natural selection-- causes the evolutionary processes to push forward.
12. Prokaryotes reproduce asexually by simple cell division. A complete reproduction cycle starts with a lag phase followed by an exponential replication phase followed by a stationary phase followed by the death phase.
13. The equation that describes the exponential growth phase is $P = 2^n$. Where P equals the number of individual bacteria in a given generation and n equals the number of generations.
14. Except in the guts of warm-blooded animals and man, exponential growth cannot be maintained for prolonged periods of time because of several factors, not the least of which is food. If it was possible for prokaryotes to replicate exponentially ad infinitum, the total weight of all the bacterial populace would soon exceed the weight of the earth. Other factors include the limitations from accumulation of waste products in the environment, and decreased bacterial reserve.
15. If the first prokaryote was produced by chance physical processes, the spread of bacteria from the point of evolutionary origin would therefore follow repeated cycles of the four phases; lag, exponential, stationary, and death. These would occur in multiple bacterial colonies as bacteria spread out from the postulated point of origin across the face of the earth.
16. Evolutionary advancement from one form into another depends upon beneficial mutation changes in the DNA. In this case, a trial was attempted to find a way for beneficial mutations to forge ahead from a less complex organism to a more complex one. If evolutionary theory is true then there must be a way for a prokaryote to evolve into a single-celled eukaryote.
17. The mutations available to advance a prokaryote to a single-celled eukaryote are the point or base pair mutation, the frame-shift mutation, the stem and loop mutation. All of these kinds of mutations must have at least a small percentage of beneficial types. Other kinds of mutations are for the most part confined in sexually reproducing eukaryotes. Until the

first pair of sexually reproducing eukaryotes came on the evolutionary scene, evolutionary processes would have to be confined to those aforementioned.

18. Point or base pair mutations only change the genetic message but adds no base pairs to the genome. Because stem and loop mutations randomly can add or subtract portions of the genetic message on a 50/50 basis, no real progress can be made in adding length to the genome because half the time with this mutation, base pairs are added and half the time base pairs are subtracted. They can, however, be involved in change.
19. Transposable element mutations are divided into two categories, insertion sequences and transposons. These mutations can be conservative or replicative. The former transposes or transfers a segment of DNA from one place and reinserts it at another place in the chromosome. Sometimes this mutation replicates the segment to be transposed. One copy is left at the donor site and the second copy is taken to another place in the chromosome where it is reinserted. This type of mutation can obviously add base pairs to the total number composing a given genome. The problem is that no new message is created; it is only repeated. The transposed segment would need to have this DNA rearranged by multiple beneficial mutations, probably by the point or base pair kind.
20. One of the easiest ways to visualize how base pairs could be added to the genome of a prokaryote is through the process of transformation. This is where some naked DNA fragments left over from the death phase of a previous bacterial colony become available in the liquid environment via lyses of bacteria. Strands of this naked DNA are then absorbed by a live bacterium which has been incorporated into its circular chromosome. After this occurs, it is conceivable that point or base pair mutations could modify the new DNA into new genes.
21. Because the death phase of bacteria is also exponential, many potentially beneficial mutations that might have occurred during exponential growth are eliminated.
22. All mutations are random both as to timing that is to when they happen and placement in the DNA, as to where they happen. Since the omnipresent judge known as natural selection has no intelligence, memory or goal, the chances for progress to be made requiring millions of new DNA messages to be added to the genome and thousands of rearrangements of the old messages of the existing DNA, the possibility of this occurring approaches as close to zero as you can get. It is impossible to visualize how advancement from prokaryote to single-celled eukaryote can occur in 2 billion years.
23. Attempts have been made in this chapter to demonstrate that it would take random evolutionary processes longer than 2 billion years to evolve a single-celled eukaryote from a prokaryote.
24. Several factors were pointed out, which actually would make the time for the evolutionary processes to be much longer than they are. Two billion years is not long enough for random evolutionary processes to change a prokaryote into a single-celled eukaryote.

Chapter 13

Fossils

Now we will examine the closely related topics of geology and paleontology. There are basically three kinds of rocks: igneous, metamorphic, and sedimentary. Igneous rock was molten at some time in the past. Examples are lava and granite. Metamorphic rocks are those re-formed by heat and pressure but without melting. Examples are slate and marble. Sedimentary rocks are rocks that have resulted from the gradual erosion of other kinds of rocks, followed by transport of tiny fragments called detrital grains to some other site, where resolidification occurs.[1], [2] Some ways that rocks erode include chemical breakdown, cracking from expansion and contraction from heat and cold, roots of plants, wind, water, and ice, especially moving glaciers. Another way occurs when water is absorbed into the pores of rock and freezes. Because water expands when it turns to ice, it causes small pieces of rock to break off. Later wind, water, or ice move these detrital grains to some other location where they settle in layers called beds on the bottom of some lake or ocean. These sediments then re-form into rock when clay or some other material binds these small fragments together like cement does the sand and small rocks in concrete.[3] Most sedimentary rock is deposited under water.[4] In addition, sedimentary rocks include limestone and what are called evaporites.[5] Mountains of limestone are mined as sources of cement. Evaporites are salt beds or gypsum beds found in many places. They are called evaporites because they supposedly formed when ancient saline lakes dried up, or evaporated, leaving the salt behind, thus the name evaporites. Coal can be classified as another example of a sedimentary rock because it formed when previous vegetation, such as peat, was buried under sediments placing it under pressure.[6] Sediments or sedimentary rocks cover the majority of the land masses on Earth.

Fossils are the remains of former ancient life forms that also can include burrows, tracks, and scat that were incorporated and preserved in rocks. It is obvious why fossils are not found in igneous rocks because the intense heat and pressure that occurred during their formation would have destroyed the evidence. So that leaves mainly sedimentary rocks as a place to look for fossils, although some fossils are found in metamorphic rocks.[7] Through plate tectonics, volcanic activity, or other forces, millions of years later these rock layers have been lifted up, many of them at angles so that they form mountains instead of ocean floors or lake bottoms. Obviously, the layers at the bottom were laid down first and each successive layer formed later.[8]

Paleontology is the study of fossils. Scientists who study fossils are called paleontologists. They have discovered that fossils of the simplest life forms are found in the lower layers, and as the layers piled up one on top of the other, fossils show a progression to more complex life forms, as the layers are examined from the lower ones to the upper ones. These facts have greatly enhanced the acceptance of Darwin's theory as the progression proceeds from the simple to the more complicated, just as his theory predicts. It is the progression of complex life forms from the lower layers and upward to which many scientists point as evidence for evolution being a fact rather than just a theory. Geologists have named the various layers, and paleontologists have studied and classified the fossils in them. Attempts have been made to calculate the time frames of the individual layers so they represent a temporal history of fossil life forms found in them, and also produce the age of the sedimentary rock layers themselves. Rough dating has been accomplished through radioactive methods and also by the fossils themselves.

There is no place, however, where all the layers of sedimentary rocks are found at one exposed location. This applies even to the Grand Canyon. However, in North Dakota, core samples taken from oil well drill bits have demonstrated all layers at one place.[9] If a huge imaginary drill bit could take a giant sample at this location it would form a tall column of rock showing the geologic history via the various layers stacked one on top of the other. This would be a true geologic column. Geologists can match various layers, some of which may be miles apart, by matching the fossils in a layer in one location with a layer in another. One reason not all locations show the layers, one on top of the other, in succession, is because after the sedimentary rocks have been lifted up by the aforementioned forces, they begin to erode and the upper layers are washed away. However, geologists have constructed on paper what they believe a real geological column might look like. It contains all the layers along with each layer's estimated age in millions of years along with each layer's representative fossils.

You will find the representation of the geologic column displayed in various museums of natural history and in geology books. Because the thickness of the various layers of sedimentary rock varies from place to place, usually the thickness of the various layers is not shown. However, most geologists believe that an accumulation of the various layers occurred very slowly, year by year. But there is no reason not to believe that during some years the rate, perhaps from a violent storm, or the ash from volcanic eruptions could have produced strata of much greater thickness.

One of the earliest geologists, in the first part of the nineteenth century, stated that what has happened to the Earth in the past was caused by the same forces working on the Earth at the present. He was referring to such phenomena as wind, water, ice, earthquakes, volcanoes, tidal waves, storms, and glaciers, to name most of the forces. Probably not much, if anything, was known about plate tectonics or continental drift in the early 1800s.

If one looks at the present geologic conditions to find out what happened in the past, then we should find sedimentary rock and fossils being formed today. The geologists who believe in slow but steady geologic processes are uniformitarians.[10] This group of scientists forms the majority. A smaller and less vocal group believes that many of the changes found in the geologic column were of a catastrophic origin. This group is the catastrophists.[11]

Now we need to ask ourselves a couple of questions:

1. Are sedimentary rocks of any significance being slowly formed today?
2. Are fossils being formed today, and are they being slowly buried?

Geologic Column

Age in Billions of Years	Events
	Rapid diversification of animals; plants and fungi appear; origin of humans about 2 million years ago
0.5	
	First multicellular organisms appear; early animals
1.0	
	First eukaryotes
1.5	
	Diverse and abundant bacteria
2.0	
	Oxygen forming photosynthesis begins
2.5	
	Diverse bacteria
3.0	
	First bacteria
3.5	
	Oldest rocks
4.0	
4.5	Earth forms

If sediments form at the bottom of lakes or oceans, then one can understand how fish could fossilize there, provided they were quickly buried alive or shortly after death. It is troubling to consider how air-breathing animals became buried in sedimentary rock that supposedly formed on the bottom of a lake or ocean. Anybody knows that if a tidal wave washed a herd of dinosaurs into a lake or ocean and they drowned, shortly afterward, scavengers would eat them and bacteria would destroy their carcasses. Finding their remains buried in sedimentary rock presumes that they must have been buried quickly and as in a catastrophe. One important sedimentary layer is the Cambrian, laid down about 550 million years ago. Burgess shale is in this layer, which contains many fossils showing soft parts partially preserved in three dimensions, or nearly so. It has been proposed that these fossils were buried quickly.[12]

As paleontologists ascended cliffs with fossil layers imbedded in them, they were impressed by the fact that the simplest life forms were in the lowest levels. But as they worked their way up, paleontologists encountered fossils of more complicated life forms. Obviously, they believed that the layers on the bottom had been laid down first making them the oldest. The increasing complexity of fossils found in higher levels seemed to confirm Darwin's theory. Its central feature was that life began simple, but driven by natural selection, slowly became more complex with the passage of time. Darwin had predicted that fossils of intermediate life forms would be found in abundance. These would connect older species with younger ones. But a problem developed when few if any fossils of the needed intermediates were found. This is why Gould and Eldredge came up with punctuated equilibrium, to be discussed in the next chapter.

Three conclusions about the present geological conditions seem obvious.

1. No known sedimentary rocks are being formed today.
2. Evidence for fossils as small as bacteria and as large as dinosaurs is found in sedimentary rock. It is possible to see how evidence for bacteria could fossilize in sediments accumulating slowly, but dinosaurs? If all of them had died in a short period of time, their carcasses would have decayed long before being buried by slowly accumulating sediments. The fact that many large fossils are buried at the same place can only indicate that they most likely were buried suddenly.
3. No known fossils are being formed today.

AMBER

Amber is a clear yellow-brown, semi-precious stone that formed from resin, a sticky sap, which flowed from an injured tree and later hardened into stone. Insects and other forms of life became stuck to the sticky substance when it was in a liquid form and were gradually completely engulfed in its death trap. Later, when the sap hardened, it became what is known as amber. Many of the trapped-life forms have had all of their soft parts preserved in this stone. The interesting aspect of these well-preserved fossils is that all of the life forms preserved in this way are easily identifiable because they look exactly like their living descendants of today. This is true even if the amber is as old as 90 million years or older. Insects, a frog, a lizard, a snail, and a mushroom have been found preserved in this semi-precious stone. Other interesting evidence of life preserved in this way includes a bird feather and some mammoth hair. So if evolution has been gradually progressing over the last 90 million years, why is it that the fossils preserved in amber are so easy to identify? These species have not evolved very much, if at all, even through millions of years. (For more information about fossils preserved in amber see *Scientific American,* April 1996, Vol. 274, No. 4, pages 84-91.)

The next chapter will show more evidence that the fossil record displays against Darwin's theory.

The discussion about fossils in this chapter is obviously extremely brief. However, as you will see in chapter 14, two well recognized paleontologists, Gould and Eldredge, discovered while they did their research that they could not find any significant evidence for the fossils needed to make the necessary intermediate evolutionary connections between any two fossil species. With this admission by them, no further attempts will be made to demonstrate the discontinuity in the fossil record in this chapter.

HOMOLOGY

In biological terms, homology is an introduction into comparative anatomy. Homology has to do with how similar parts or organs of different species vary in structure and function. In evolutionary terms, these differences in structure were supposed to have come about through descent with modification from a common ancestor. They also are presumed to demonstrate adaptation of various organs of similar structure to different functions.

Even though many of the examples of homology are found in organisms alive today, from an evolutionary point of view, these living forms were supposed to point backwards to a predecessor long since dead. The study of homologies in living forms is supposed to be a vicarious way of studying fossils. That is why this topic is included in this section.

The following diagram (Figure 13-2) respectively shows the forearm extremity bones of man, monkey, and bat. Many other examples of the forward extremity can be cited such as manatee, moles, and cats, just to mention three more. All have similar structures, but not all have the same function. Man and monkey use their pendactyl hands to grasp, the bat uses these similar structures to fly. The manatee uses the forward limbs to swim, the mole to dig, and the cat to walk.

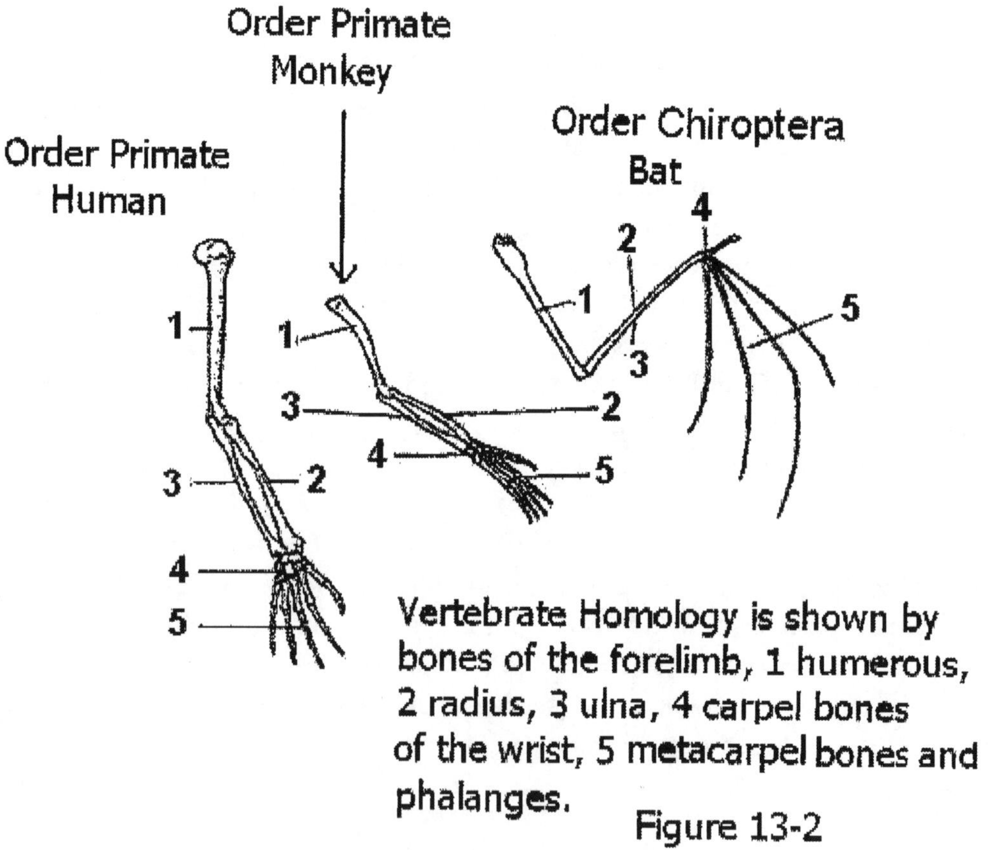

Figure 13-2

According to evolutionary theory, these similarities in structure are supposed to demonstrate that these animals have descended from a common ancestor many millions of years ago.

An article printed in 1971 by Oxford University Press, written by a Sir Gavin de Beer, is an excellent source to discuss this topic. De Beer was formerly Professor of Embryology at the University of London and had served as Director of Natural History at the British Museum. He has authored a few books on the subject of evolution.

The article written by de Beer points out many problems in homology which shows that it does not indicate common evolutionary ancestors.

Darwin wrote in his book, *The Origin of Species,* chapter 13, *"This is the most interesting department of natural history, and may be said to be its very soul. What can be more curious than the hand of man; formed for grasping, that the mole for digging, the leg of the horse, the paddle of the porpoise, and the wing of the bat, should be all constructed on the same pattern, and should include similar bones in the same relative positions?"* Darwin implied that homology formed the very soul of natural history or evolution.

One of the evolutionist's favorite examples used as evidence to show descent from ancient ancestors, is the left recurrent laryngeal nerve. By tracing its pathway through the bodies of successively higher life forms, this nerve has a pathway that is significantly different from the right laryngeal nerve. This nerve, along with its counterpart on the right, enervates the muscles of the larynx (voice box) and is used when we speak. Its pathway is much longer on the left side than on the right. The reason for this is that as the mammalian embryo develops, this nerve becomes caught behind one of the blood vessels and pulled down into the chest as the embryo grows. It has a 'U-shaped' course from the head downward into the chest and back up again to the larynx or voice box. This pathway is by necessity, then, considerably longer on the left side than the right. On the right side, the laryngeal nerve follows a pathway from the head directly to the larynx and is therefore much shorter. Scientists, who compare the anatomy of various animals, have traced the pathways of this nerve, or its counterpart, from lower animals up to mammals. They have found a correspondence that is touted as evidence of descent of mammals all the way back to amphibians. Of course, the most extreme example of this is found in the giraffe. This means that the left recurrent laryngeal nerve is quite a few feet longer than the right since it follows a pathway down from the head into the chest and back up again to the larynx. The right laryngeal nerve runs directly from the head to the larynx.

If these similarities of structure between various vertebrate animals are to demonstrate one of the major foundation stones of evolution, referred to by Darwin as its soul, then it must bare up under scrutiny; the way homology is presented by evolutionists, as a group, is that it is a form of positive circumstantial evidence. To reiterate, positive circumstantial evidence shows that something might be true, but never that it is.

In his article, de Beer describes many points of negative circumstantial evidence, which shows that something is untrue or could never have happened.

Many animals contain body segments, even the earthworm. In higher animals, vertebrae form part of the body segments. Early in embryonic development of higher animals, the skull and its divisions form part of the segments as well. Various nerves and extremities originate as derivations of the various body segments. If evolution is true, then we should expect all of the forelimbs of the vertebrates; similar parts (that is the homologous bones of the arm, elbow, forearm, wrist) ending in five digits to derive from the same body segments. Well, they don't. Here is a quote from de Beer himself, "*So in the newt the forelimb is formed from trunk segments 2, 3, 4, and 5; in the lizard 6, 7, 8, and 9; in man from trunk segments of 13 to 18 inclusive.*" If all the homologous parts of these animals have derived from ancient ancestors, then these parts should have derived from the same body segments, but they do not. The same phenomenon is noted in the development of the occipital arch of the skull. In animals starting with the ancient shark on up to frogs, newts, reptiles, and mammals, the origin of this homologous structure derives from different body segments except that reptile and mammal occipital arches derive from the same segments.

De Beer lists several other discrepancies of how homologies don't support the evolutionary theory. To quote him again, "*Since every organ and structure in any organism has come into existence only as a result of embryonic development, it is natural to look to embryology for evidence on homologous structures.*" Alas, de Beer seems to say, the evidence shows that many homologous parts of different animals are derived from different embryonic tissue and from different locations on the egg.

De Beer almost seems to figuratively throw his hands up in dismay as he emphasizes by italics in his article the following quote. "*Therefore, correspondence between homologous structures cannot be pressed back to similarity of position of the cells of the embryo or the parts of the egg out of which these structures are ultimately differentiated.*"_

After the egg is fertilized and begins to divide, de Beer points out that the cells soon differentiate into three distinct germ layers: the ectoderm, the endoderm, and mesoderm. Generally speaking, the ectoderm goes on to differentiate further into the epidermis, the nervous system, the sense organs, and nephridia, a form of primitive kidney. The endoderm continues to change into gastro-intestinal tract from which derive (in vertebrates) the thyroid, lungs, liver, pancreas, appendix, and urinary bladder. From the mesoderm develops the dermis (or under layer of the skin), connective tissue, cartilage, bone, muscles, germ cells, genital ducts, and kidneys, if the nephridia has been discarded.

The previous generalization very soon became biological dogma that homologous organs must always arise from the "correct" germ layer. This dogma, as de Beer calls it, soon became shaken when various experiments performed on newt embryos showed that when the neural crest was removed surgically, and as the embryo continued to grow, the cartilage of the jaws was also missing. Since the neural crest was supposed to only change into nerve cells (which in turn had developed from ectoderm), and since cartilage was supposed to only have developed from mesoderm, de Beer considered this new discovery to be "morphological heresy." The final insult, to quote de Beer, "*So the imagined embryological specific monopoly of the germ layers and what they invariably give rise to was shattered ... It is therefore necessary to give the lie directly to the entry on 'Homolog' in the glossary by W.S. Dallas which Darwin most unfortunately appended to his 6th edition of the Origin of Species. It defines homology as 'That relation between parts which results from their development from corresponding embryonic parts.'"*

After this short quotation from the Origin, de Beer continues emphatically stating that, *"This is just what homology is not."*

In 1971, for de Beer to claim that a portion of Darwin's *The Origin of Species,* was wrong, would be similar to Copernicus's claim in the 1540s that the earth orbits the sun rather than vice versa, as Ptolemy and others Greek astronomers claimed. De Beer's descriptions were made several decades ago. Yet the scientific community has declined to accept these findings. The high school biology text, *Biology, Visualizing Life* by Johnson, copyrighted in 1994, shows homologous structures as proof of evolution.

If this were not enough, de Beer waits until the last to explain the most damaging evidence of all against Darwin's theory, and this he refers to as "the *worst* shock of all." This area concerns the genes. Since genes are the ultimate source of storage of information for the development of a given characteristic found in a phenotype (that is a given organ or part of an organ), then we should only expect that these genes had their source from common ancestors and had been preserved by natural selection with modification. De Beer points out that this is not the case. He proceeds to give several examples to prove his point. What this means is that characteristics controlled by identical genes are not necessarily homologous: "*Therefore, homologous structures need not be controlled by identical genes, and homology of phenotype does not imply similarity of genotypes.... It is now clear that the pride with which it was assumed that the inheritance of homologous structures from a common ancestor explained homology, was misplaced; such inheritance cannot be ascribed to identity of genes."*

Since the inference that homologous structures imply descent from a common ancestor has been proven untrue, the claim of Darwin that homology forms the foundation for his evolutionary theory that he referred to as "its very soul" has been jerked out from under it. Homology, as Darwin interpreted it, has been proven untrue and must be abandoned by science as evidence for the truthfulness of evolutionary theory. If anything, it is a source of negative circumstantial evidence that proves that something either could not happen or did not happen.

SUMMARY

1. There are three kinds of rock: igneous, metamorphic, and sedimentary.
2. Igneous rock, like lava or granite, was molten at sometime in the past.
3. Metamorphic rock is re-formed rock from heat and pressure, but not melted.
4. When any type of rock is broken into small detrital grains, resulting from such physical forces as water freezing in pores, wind, ice, etc., and these grains are transported to some other place by wind, water, or glaciers, etc., they form layers that later become resolidified into beds known as sedimentary rock.

5. Fossils, the remains of previously living organisms, are found almost exclusively in sedimentary rock, although some may be found in metamorphic rock. None are found in igneous rock.
6. Sedimentary rock almost always was formed under water in layers called beds.
7. Later, due to actions of plate tectonics, these layers have been lifted up, some at angles and may become mountains.
8. Obviously, layers on the bottom were laid down first with all successive layers being laid down later.
9. The fossils of simpler life forms are found in the lowest layers and, as more layers were laid down in succession, the fossils of more complicated life forms are found preserved in them.
10. Sedimentary rock, or sediments that have not reformed into rock, cover the majority of the Earth's surface.
11. There are a few places on Earth, at the present time, where all of the sedimentary layers or beds are piled on each other at the same location. Core samples from oil well drilling bits in North Dakota, are examples.
12. If a giant drilling machine could cut a continuous core of rock layers at one of these places, it would form a geological column. Because this is impossible, geologists constructed on paper, what they believe an imaginary geological column might look like.
13. Not all sedimentary rock layers are found even at the Grand Canyon where nearly a mile deep trench of sedimentary layers is exposed.
14. By using radioactive clocks and the fossils themselves, time estimates are applied to the various layers giving rough estimates of their ages. The thickness of the various layers can be divided by the length of time in years that these layers were forming. This gives the average thickness of sediments laid down each year, as being usually only a few millimeters.
15. Obviously fossils had to be buried quickly or they would have decayed. Therefore, there had to be times when the rate of accumulation of sediments was exaggerated in order to bury the fossils quickly and preserve them in rock.
16. The geological column and fossils do give us a somewhat obtunded idea of this planet's natural history.
17. No known sedimentary rocks are being formed today.
18. No known fossils are being formed today.
19. Because fossils of the simplest life forms are found in the lowest sedimentary rock beds and because fossils with more complicated life forms are found as layers of the geologic column were successively laid down, this finding has given evolutionary theory a great impetus as this is exactly what the theory predicts. To this phenomenon, evolutionist point as proof that evolution has occurred.

20. Fossils preserved in amber millions of years ago look the same as their descendants do today. They have not evolved very much, if any, because the life forms preserved in the amber are basically identical to their living counterparts today.
21. Homologous structures, such as the pendactyl forelimb of various vertebrates, which Darwin touted as evidence that many animals had descended modified from a common ancestor, cannot be traced back to their origin from the same body segment, place on the egg, or even the same gene. Therefore, homology cannot be used as evidence of descent from a common ancestor, because it means that natural selection would have invented the same pendactyl configuration many times, all by chance with no intelligent memory, or goal.

Chapter 14

Confounding Boundaries

With all of the evidence presented thus far against evolution being able to produce complex life in the time available, does the fossil record actually show that it did occur? Is the negative circumstantial evidence outweighed by unequivocal positive circumstantial evidence that proves evolution really did take place? In the prologue to his book, *Reinventing Darwin,* published in 1995, Niles Eldredge writes that there is a continuing debate between two factions of evolutionary biologists. He indicates that this is a High Table of evolutionary debate. He says, "*In British colleges, dining rooms traditionally feature a High Table at the head of the room, where the elite of the institutional staff, the doyen, sit aloof from the student hoi polloi.*" (page ix) Although he hastens to point out that evolution *per se* belongs to no individual, he also proposes that there are names in the forefront of evolutionary theory who are hypothetically sitting across from each other at this imaginary High Table of evolutionary debate. Indeed, there is a debate going on among the elite evolutionary gurus whose names frequently pop up in evolutionary literature. Eldredge quickly gives names to the two sides of the debate, calling one side the Naturalists and the other side the Ultra-Darwinists. In the Ultra-Darwinists group, he names George Williams, Richard Dawkins, and John Maynard Smith. In the former group, he names himself, Stephen Jay Gould (before he died), and Elisabeth S. Vrba. Of course, there are many unidentified others on each side.

At this hypothetical evolutionary table, the biologists are not seated for dining on choice gourmet delights, but are there so that they can argue about the problems involving the validity of the Neo-Darwinian synthesis. From nearly all biology textbooks, one would never suspect that there is a debate in progress. We have been led to believe that the Neo-Darwinian synthesis is so well-established that it has moved from theory almost to law. *Reinventing Darwin* is an eye-opener on that score.

(Unless otherwise noted, the following quotations are from *Reinventing Darwin*, by Niles Eldredge, 1995, John Wiley & Sons, Inc. New York, NY).

Right away, Eldredge points out, "*What is really at stake is diametrically opposed suppositions of how evolutionary biology should be conducted.*" (page 2) Later, he accuses the Ultra-Darwinists of suffering from myopia, a scientific term for near-sightedness. Of all things, Eldredge then accuses his opponents of having a *"slavish adherence to tradition that dates back to Darwin."* (page 3) Three sentences summarize the main Ultra-Darwinian tenants just as Darwin theorized. He noted the very slight differences between each succeeding generation of any species and the next. He proposed that natural selection would preserve even the slightest beneficial differences that he noted between succeeding generations and would not allow deleterious changes to survive. From this point on, the gradual changes that take place between succeeding generations need only be extrapolated over eons of time to produce the huge variety of biota seen today.

Eldredge says this: *"Simple extrapolation does not work. I found that out back in the 1960s, as I tried in vain to document the examples of the kind of slow, steady directional change we all thought ought to be there, ever since Darwin told us that natural selection should leave precisely such a telltale signal as we collect our fossils up cliff faces. I found instead, that once species appear in the fossil record, they tend not to change very much at all. Species remain imperturbably, implacably resistant to change as a matter of course - often for millions of years."* (page 3) He referred to the 1972 paper co-authored by himself and Gould in which they called the phenomenon of the stability of species as "stasis." (page 3) After millions of years of stasis, new species suddenly appear in the fossil record with apparently no ancestral past. So where did they come from? The paleontologist found no intermediate fossils to connect the old fossils and the younger ones. There were big gaps and big jumps between one form of biota and the next. This phenomenon in the fossil record is referred to as discontinuity.

Eldredge points out that generally the Ultra-Darwinists are geneticists and that the Naturalists, for the most part, are paleontologists. The basis for the debate between these two eminent groups of evolutionary scientists hinges on the fact that the geneticists can see no way for sudden large changes to take place because of the restrictions imposed by genetic rules coupled with random mutations and natural selection. The paleontologists found that discontinuity is universal in the fossil record, with intermediate forms absent. Stasis is the norm. Once a given species appears in the fossil record, it remains the same for millions of years. The paleontologists' findings of new biota suddenly making their appearance in a given stratum without any ancestral past, proved to be a big surprise. In chapter 9 of *The Origin of Species*, Darwin said, *"By the theory of natural selection all living species have been connected with the parent-species of each genus, by differences not greater than we see between the varieties of the same species at the present day; and these parent-species, now generally extinct, have in their turn been similarly connected with more ancient species; and so on backwards, always converging to the common ancestor of each great class. So that the number of intermediate and transitional links, between all living and extinct species, must have been inconceivably great. But assuredly, if this theory be true, such have lived upon this earth."* But in the strata of sedimentary rocks, not only did the new forms suddenly appear, but they also existed for millions of years, virtually unchanged, before they passed off the scene in extinction. This lack of progression of biota to evolve (stasis) and the conspicuous absence of intermediate fossils, according to Darwin, would prove his theory untrue. But, the fossil record does show a progression of organisms from simple to more complex, as succeeding upward sedimentary layers are examined. However, there are no intermediates to explain this progression, which Darwin predicted would be found and which his theory demands.

The obvious problem that paleontologists have to face, with respect to Darwin's predictions, is the one of discontinuity. **There is a conspicuous absence in the fossil record of the almost innumerable number of intermediates needed to bridge the gaps between a simpler life form and a more complicated one.** In Neo-Darwinian terms, at least hundreds or even thousands of intermediate forms would be needed between the so-called divergences of one class of biota into another, such as the transition of fish to amphibian. But what the paleontologists find is the sudden appearance of new biota with no hint of the many intermediates needed in their ancestral past to explain their appearance in evolutionary terms. Then there are millions of years of stasis with no significant change. Both of these findings in the fossil record challenge the Neo-Darwinian synthesis. The validity of the finding of stasis in the fossil record as described by Gould and Eldredge has been corroborated by a report in the December 5, 2003, issue of *Science* magazine (page 1645). An ostracode fossil was found with most of its soft tissues well preserved in three-dimensions. It demonstrates no significant change in 420 million years with present-day descendants.

Both sides of this evolutionary debate face a quandary. Each is sitting on the horns of a dilemma. On one horn sit the Ultra-Darwinists who predict that intermediates must be present in the fossil record and abhor the thought of stasis. Their position is based on the realization that rules of genetics prohibit sudden jumps from simple to more complex life forms. On the other horn sit the Naturalists, who found that intermediates are conspicuously absent in the fossil record and that stasis is the rule. Sudden jumps are implied when the paleontologists find a new species among the fossils that suddenly appeared out of nowhere. Therefore, it is easy to see something has gone afoul with the Darwinian prediction of a very slow transition from one form to another and the sudden jumps, as observed in the fossil record. This implies saltations (or big jumps) which genetic laws prohibit.

Eldredge and Gould proposed a possible way around this conundrum in 1972 in their now-famous paper regarding fast evolution called punctuated equilibrium. Eldredge hastens to deny that punctuated equilibrium implies saltations, which Darwin stated in *The Origin*, could not exist. Eldredge believes that evolution happens very fast in some situations (sudden jumps). This is why he thinks that this process leaves very few, if any, intermediates in the fossil record.

You can see that there is a crack, if not an outright rift, that has developed between these two groups of evolutionary scientists about Darwin's theory. Practically all high school and college biology textbooks don't discuss this. They only give a unified picture of consensus among scientists of note that the theory of evolution has been proven and that it is nailed down and there are no debates.

According to classical Darwinian theory, species are supposed to be only temporary forms of life, each of which is constantly in a state of flux or change from generation to generation. Two quotations from Eldredge's *Reinventing Darwin* can put this into proper perspective.

"Thus our term 'phyletic gradualism' in general means 'slow, steady change by degrees.' In particular, it refers to the slow, steady transformation of an entire species. We presented evidence that, contrary to the long-held picture of gradual evolutionary change through time, most species hardly change much at all once they appear in the fossil record the phenomenon we call 'stasis.' We pointed out that paleontologists clung to the myth of gradual adaptive transformation even in face of plain evidence to the contrary paleontology's 'trade secret,' as Gould later called it.' (page 63)

So why did the paleontologists have a trade secret? Why didn't they report these findings to the various societies of science and the general public at large? Eldredge gives the answer.

"For the most part, it has been paleontological reluctance to cross swords with Darwinian tradition that accounts for the failure to inject the empirical reality of stasis into the evolutionary picture." (page 68)

So here we see a group of twentieth century scientists who were afraid to tell what they knew to be true from their research because they did not want to go against the nineteenth century scientist, Darwin, who had invented a paradigm that they thought took priority over reality.

This has happened before. Other groups of scientists have continued to agree with what was postulated by preceding smart men, despite observed, conflicting data. Aristotle's and Ptolemy's teaching of the Earth-centered universe went unchallenged until Copernicus and Galileo. If those in authority who tried Galileo for his so-called scientific heresy had taken the trouble to look through Galileo's telescope to see the mountains on the moon and the phases of Venus, they would have seen that the heavenly bodies beyond the Earth were not perfectly smooth as envisioned by Aristotle. Also, the phases of Venus helped to prove that the sun was the center of our solar system and not the Earth. When Kepler was plotting the planetary orbit of Mars, he had difficulty believing that it followed an ellipse, rather than a perfect circle as Aristotle had also implied. Einstein was reluctant to believe that the universe was expanding, even though his equations indicated that it was. He thought that he would be going against the great Isaac Newton, who thought that the universe was infinite and static. So, Einstein added his cosmological constant to keep the universe static to comply with Newton, until Hubble's research proved otherwise. These are just a few examples of the reluctance of succeeding generations of scientists who were afraid to confront a paradigm that had become dogma.

Even Gould and Eldredge were reluctant to tackle head-on the Neo-Darwinian synthesis. Their idea of punctuated equilibrium is nothing more than a biological "cosmological constant," if you will. They postulate that evolution happens quickly after a major change in the ecosystems where habitats are destroyed by catastrophic events such as hurricanes, volcanoes, earthquakes, etc. When ecosystems are destroyed, geneticists or gradualists (the Ultra-Darwinists) predict that extinction of the biota will take place. The naturalists predict that species will seek out new habitats or evolve very quickly, thereby leaving so few intermediates that they have never been found as a sort of "now you see them, now you don't" idea.

Stasis and total lack of intermediate life forms in the fossil record prove that gradualism envisioned by Darwin just plain doesn't exist. The idea for one class of organisms to evolve into another by following the tenants of punctuated equilibrium is preposterous because of the difficulties encountered from genetic rules, which if evolutionary theory were true would require slow changes. The fossil record, instead of confirming the evolutionary theory, provides more evidence against it.

So what did Darwin have to say on this very subject? In *The Origin of Species* in the Chapter 6, titled "Difficulties on Theory," Darwin says, "...*natural selection can act only by taking advantage of slight successive variations; she can never take a leap, but must advance by the shortest and slowest steps."* Then in chapter 9, titled "On the Imperfection of the Geological Record," Darwin predicts what the geological findings should be: *"But just in proportion as this process of extermination has acted on an enormous scale, so must the number of intermediate varieties, which have formerly existed on the earth, be truly enormous. Why then is not every geological formation and every stratum full of such intermediate links? Geology assuredly does not reveal any such fine graduated organic chain; and this, perhaps, is the most obvious and gravest objection which can be urged against my theory. The explanation lies, as I believe, in the extreme imperfection of the geological record."*

Then on page 95, Eldredge proceeds to comment on what Darwin had thought. *"From an evolutionary point of view, then, the fossil record has long had two strikes against it: its gappiness, and its uncertainties about where its fossilized animals and plants might have come from. Darwin devoted two chapters of the Origin to geological time and the fossil record.* In the first (Chapter 9, 'On the Imperfection of the Geological Record') *he established the enormity of geological time, thus providing himself all the time necessary for life to evolve, and making a great contribution to our understanding of earth history. In that same chapter, Darwin was at pains to explain why paleontologists had not as yet found numerous examples of what he called 'insensibly graded series' of fossils. Why, in other words, had paleontologists not been able to document the very pattern Darwin thought evolution must leave in the rocks?"*

"For one thing, Darwin mused, paleontology in his day was still in its infancy. Surely, he wrote, paleontology would eventually provide full corroboration of his theory. Darwin actually believed that his entire theory of 'transmutation' (or, 'descent with modification' he never called it 'evolution' in the Origin) would stand or fall on the eventual recovery of many examples of gradual evolution in the fossil record."

"No wonder paleontologists shied away from evolution for so long. It seems never to happen. Assiduous collecting up cliff faces yields zigzags, minor oscillations, and the very occasional slight accumulation of change - over millions of years, at a rate too slow to really account for all the prodigious change that has occurred in evolutionary history. When we do see the introduction of evolutionary novelty, it usually shows up with a bang, and often with no firm evidence that the organisms did not evolve elsewhere! Evolution cannot forever be going on someplace else. Yet that's how the fossil record has struck many a forlorn paleontologist looking to learn something about evolution." (page 95.)

So how does Eldredge try to get around this conundrum? Again he falls back onto punctuated equilibrium as an explanation on page 66.

"*Naturalists, on the other hand, say that the most likely response of a species to environmental change is habitat tracking. The second most likely response to environmental change is extinction, which generally follows when suitable habitat cannot be found. The least likely outcome is wholesale, linear transformation of an entire species to meet the new environmental exigencies.*"

What this means, according to the naturalists, is that if a given species' habitat has been drastically changed by some catastrophic event, the species will move someplace else, which Eldredge calls "*habitat tracking.*" When these species die in this new place, their fossils obviously would seem to appear "suddenly" in the fossil record at that location. The second most likely possibility is extinction, and the third is the magical sudden transformation of fast-track evolution that leaves very few, if any, fossils.

On page 74: "*As has become abundantly clear, many sudden anatomical shifts in the fossil record reflect not evolution, but migration from elsewhere of related, but different stocks.*"

This only shows that the sudden appearance of different fossils reflects a migration of an already established species from someplace else and not the evolution of any new species.

On Page 66: "*Traditional evolutionists, including latter day ultra-Darwinians seated at the High Table, have it the other way around: environmental change begets evolutionary transformation through natural selection; failing that, we expect extinction. Habitat tracking doesn't even enter into the ultra-Darwinian picture except to be airily dismissed as 'fable,' as George Williams has done in his recent Natural Selection: Domains, Levels and Challenges (1992).*"

Eldredge goes on to try to explain in different evolutionary terms what Darwin had predicted. He says that the reason suddenly new, and more complicated forms interrupt millions of years of stasis without any intermediates found in the fossil record, is that it is difficult for new speciation to succeed in a given ecological niche while trying to compete with all the other plants and animals in that niche.

On page 151: "*That theory is, once again, ably supplied through the now familiar notion of species sorting. Once again, we must consider fledgling species. In a world suddenly devoid of a majority of the species that used to staff the ecosystems, we can imagine the probable fate of the few new species that might, in the course of time, appear through normal speciation. They would almost certainly have a far better chance of becoming established in a newly under-populated world than would be the case in more 'normal' times. If not quite a case of 'anything goes,' surely the exigencies normally facing fledgling species would be greatly relaxed after large-scale extinctions. Survival rates would be far higher than usual. It is the rate, not of speciation per se, but of successful speciation, that goes way up after a major extinction event.*"

On page 151: "*It is equally abundantly clear that no manner of old-style, simplistic ultra-Darwinian extrapolation can be of any use in addressing these major patterns of evolutionary history.*"

page 153: *"The vast majority of species that have ever lived have not only become extinct, they also have left no descendants."*

page 154: "*...little or no evolution occurs unless and until an extinction event occurs to shake up entrenched ecosystems."*

Notice that Eldredge claims most species that have ever lived haven't left any descendants. They have become extinct. Only the fossils left behind determine the knowledge of a species' prior existence. So how does Eldredge know that these species ever existed? He also is saying that only minimal evolution occurs (that it is very difficult for a new species to get a foothold) unless extinction caused by some catastrophic event kills off the established ecosystems so that the new species can get a foothold without competition.

Darwin taught that competition between various living organisms was supposedly the driving force that perfected living organisms and that yielded evolutionary progression, selected over eons by natural selection! But Eldredge seems to abandon another one of Darwin's principal pillars: natural selection or survival of the fittest (which is supposed to hone succeeding generations to perfection through competition). Eldredge believes that extinction from catastrophic causes removes the competition so that his postulated new life-forms can become established with less or no competition.

page 99: *"What we were saying is that evolution looks instantaneous in the fossil record, but is not. Indeed some evolutionary geneticists have said that the 'five to fifty thousand years' estimate is, if anything, overly generous. Speciation events may often require even less time to take place."*

Again, Eldredge says that speciation events happen rather quickly. If true, why are the biota preserved in amber, so easy to recognize? Surely, over 90 million years there should have been some recognizable changes.

page 156: *Evolutionary history, then, is deeply and richly contingent. Gone are the last vestiges of the idea that evolution inevitably and inexorably replaces the old and comparatively inferior with superior new models. Evolution, at least on a grand scale, is not forever tinkering, trying to come up with a better mousetrap."*

"It's the other way around: species, and the ecosystems that their component organisms staff, are tenacious. They "work" perfectly well and, once entrenched, are unlikely either to change or to be displaced by newly evolved taxa – unless and until extinction knocks ecosystems off their tracks. Then evolution breaks loose."

Right away, Eldredge brings up the accusations that the Ultra-Darwinists make against the rapid leaps from one species to the next, which punctuated equilibrium espouses. This is a very valid accusation. How can evolution break loose real fast and traverse the sea of chaos that exists between one island of biota to the next, using the trial-and-error method with no intelligence, memory or goal? For punctuated equilibrium to explain why the intermediate fossils are absent, rapid leaps would be required, which leave practically no fossils to prove that rapid leaps ever happened. To defend his position, he draws on the example of the plant clarkia. Scientist Harlan Lewis discovered in the 1960s, that some populations of clarkia had a fairly severe mutation, which occurred with a rather high rate of frequency almost annually. Because these mutated forms could not pollinate other clarkia plants from whence they had arisen or vice versa, they were, in fact, a new species. They were "*reproductively disjunct.*" They did not flourish. Each year when a new group of mutated forms appeared, these small mutated populations became extinct. Eldredge reasoned that they could not compete against their parent stock and, therefore, were "swamped out" of existence. On page 122, Eldredge quickly postulates that the mutated clarkia would have survived better in their ecological niche if they had not had to compete with the parent stock.

We may be justified in asking, was it not the severe mutation, rather than the competition from the parent stock, that caused their demise? If the mutation had been beneficial, the mutated form should have crowded out the parent stock. Even if the mutated clarkia could have survived and left fossils for some scientist to find millions of years later, he would still be able to tell that they were clarkia. This not only would have shown that this species danced around a sort of median, but also it would not have proved either the punctuated-equilibrium scenario or the Ultra-Darwinism position.

It is interesting that the late Stephen Jay Gould, in his book *Dinosaur in a Haystack*, talks about this very thing on pages 405-407. You will remember that it was Gould who, along with Niles Eldredge, proposed punctuated equilibrium in 1972. Some perfectly preserved fossil leaves of clarkia were discovered in rocks where they were buried 20 million years ago. E. M. Golenberg and colleagues compared 820 base pairs of DNA from this ancient clarkia source with an 820 base pair sequence of DNA in a closely related living species. They found only 17 base pair substitutions or differences between the fossil leaf and living species. Of the 17, there were 13 silent substitutions and four neutral substitutions, which coded four different amino acids. Obviously, the four neutral changes in amino-acid sequences had to be amino acids that functioned similarly to the ones replaced or had occurred in non-critical positions in the resulting protein molecule. Otherwise natural selection would have discarded the living species. The important point is; only minuscule changes occurred in 20 million years. No big jumps. The original published source from which Gould obtained his information was the April 12, 1990, issue of "*Nature*."

Eldredge believes that there is a "built-in bias" for what he calls fledgling species to survive and repopulate devastated ecological areas. Supposedly this new theory explains the lack of intermediate fossils and stasis in one fell swoop.

What Eldredge is implying here is another extrapolation. He draws a principle from the study of populations of modern clarkia that, for some reason, produced "*fairly severe mutations*" and that all became extinct because their predecessors supposedly swamped them out. However, it could easily have been the severe mutations themselves that caused their demise. The mutations were obviously not beneficial. Despite this, Eldredge tries to show us that new emerging species can only get a foothold when there is a severe disruption in their habitat. This way, he claims, by his reinvention scenario of punctuated equilibrium, the old species destroyed by some catastrophic event no longer would be able to compete and swamp out the new species.

In other words, according to Eldredge, with a severe disruption in the habitat, evolution breaks loose. New species develop quickly and leave very few fossils thereby explaining the lack of intermediates in the fossil record.

If evolution breaks loose when ecosystems are disrupted, then ecosystems would have to be destroyed every few generations for these supposedly new life-forms to evolve rapidly. Do major catastrophes such as earthquakes, volcanic eruptions, avalanches, fires, floods, or collisions from asteroids happen that often? The onset and rapid progress of complex life forms during the Cambrian explosion cannot be explained by punctuated equilibrium. Since the Cambrian explosion, there have been at least one or two major extinctions probably brought on by crashes to the Earth's surface of falling celestial bodies that obviously destroyed multiple ecosystems. However, since the Cambrian explosion, no new phyla have emerged. Why not? Are we supposed to believe that new species suddenly appear due to rapid evolutionary progress, leaving no telltale signs of intermediate life forms just because an ecosystem was destroyed?

In living forms, the transitional or intermediates needed as evolutionary ancestors between classes are uniformly absent. Paleontologists researching the fossil record have been unable to find transitional forms or intermediates needed among these dead specimens. This is very strong negative circumstantial evidence against Darwin's theory, because his theory would require that there be hundreds, or even thousands, of intermediates between the various classes. They are conspicuously absent. Pure random chance with outcomes augmented with choices of natural selection, would not allow all these intermediates to be missing (punctuated equilibrium, not withstanding) if, in fact, they had occurred.

If the fossil record lacks any significant examples of the multiplicity of intermediates needed to demonstrate descent with modification, then the branches shown in the evolutionary-tree diagrams do not unite at the trunk. If they do not unite, then there can be no common ancestors. If evolution were true, then we should find a continuum of intermediate fossils and living forms nearly everywhere we look. Even Darwin, in his letter to a friend, said, *"Imagination must fill up the very wide blanks."*[1] There are obvious gaps between organisms of the same class that have similar characteristics; gross differences exist between organisms of two separate classes.

Creation on the other hand contains reasons why intermediate life forms are missing from any living species and largely from fossils. According to Genesis 1, the simplest life forms were created on Genesis day three, followed with more complex life forms on Genesis days five, and six. None had evolved. Therefore, no intermediates would be needed to connect the classes of biota. Apparently from the outset, life cycles of the simplest life forms created on days five and six included death. Their fossils would have been buried in successive ascending layers, as eons of cosmic time passed. If so, mankind knew what God meant by death. Originally, it seems that only humans were granted everlasting life, based on their continued belief in God and eating fruit from the tree of life. But instead of believing God, who told them they would die if they ate the forbidden fruit, they chose to believe the serpent, who said they wouldn't die and they'd be like God. Their disbelief and the deprivation from tree of life fruit, brought mankind death.

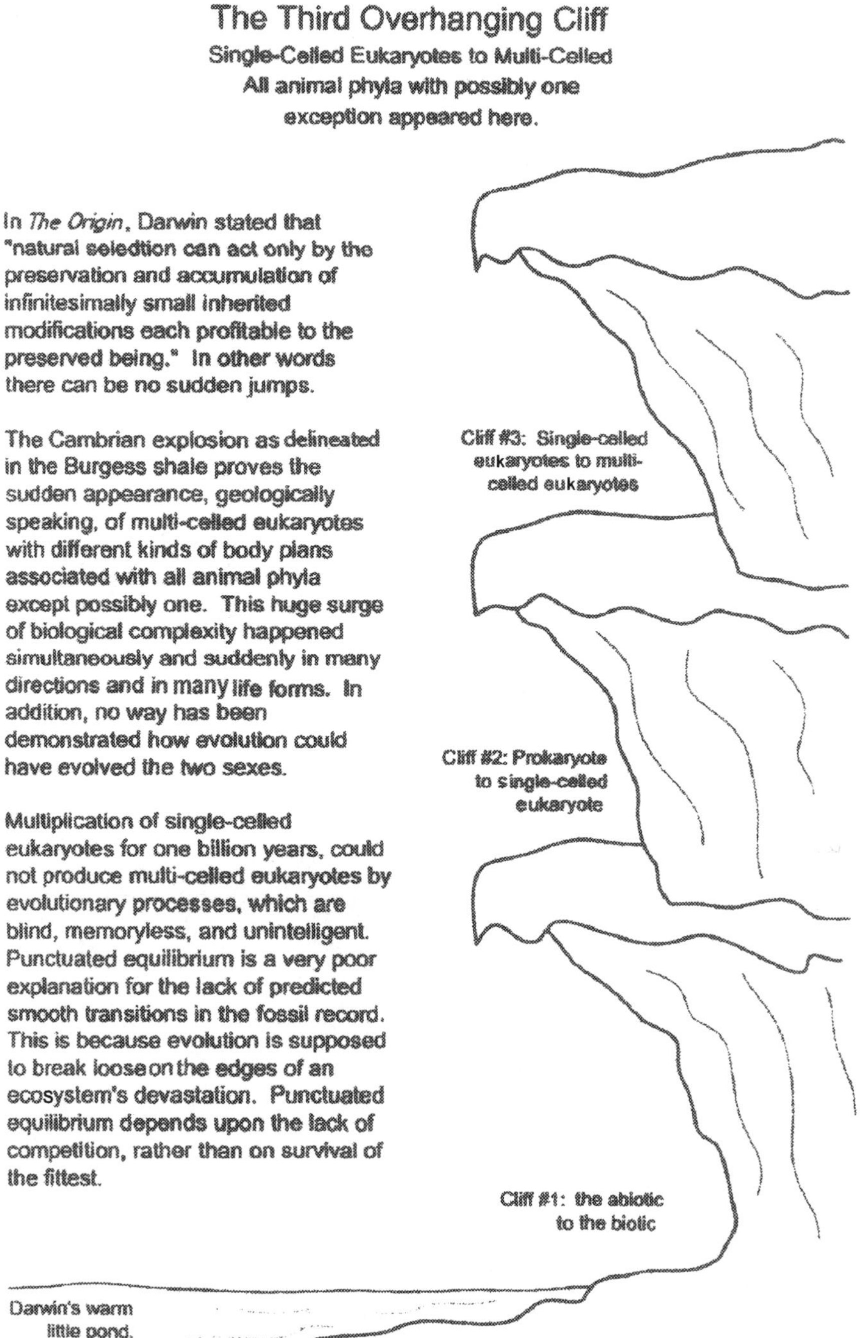

The Third Overhanging Cliff

Single-Celled Eukaryotes to Multi-Celled

All animal phyla with possibly one exception appeared here.

In *The Origin*, Darwin stated that "natural seledtion can act only by the preservation and accumulation of infinitesimally small inherited modifications each profitable to the preserved being." In other words there can be no sudden jumps.

The Cambrian explosion as delineated in the Burgess shale proves the sudden appearance, geologically speaking, of multi-celled eukaryotes with different kinds of body plans associated with all animal phyla except possibly one. This huge surge of biological complexity happened simultaneously and suddenly in many directions and in many life forms. In addition, no way has been demonstrated how evolution could have evolved the two sexes.

Multiplication of single-celled eukaryotes for one billion years, could not produce multi-celled eukaryotes by evolutionary processes, which are blind, memoryless, and unintelligent. Punctuated equilibrium is a very poor explanation for the lack of predicted smooth transitions in the fossil record. This is because evolution is supposed to break loose on the edges of an ecosystem's devastation. Punctuated equilibrium depends upon the lack of competition, rather than on survival of the fittest.

Cliff #3: Single-celled eukaryotes to multi-celled eukaryotes

Cliff #2: Prokaryote to single-celled eukaryote

Cliff #1: the abiotic to the biotic

Darwin's warm little pond.

Figure 14-1

It was Darwin in *The Origin* Chapter 9 who said: *"By the theory of natural selection all living species have been connected with the parent-species of each genus, by differences not greater than we see between the varieties of the same species at the present day; and these parent-species, now generally extinct, have in their turn been similarly connected with more ancient species; and so on backwards, always converging to the common ancestor of each great class. So that the number of intermediate and transitional links, between all living and extinct species, must have been inconceivably great. But assuredly, if this theory be true, such have lived upon this earth."*

Darwin, himself, in this quotation from *The Origin,* gives the litmus test to prove, or disprove, the validity of his theory. He states that if this theory is true, there must have lived upon Earth, "inconceivably great" numbers of intermediates to connect present living forms with the past living forms. Darwin may take the liberty of filling in the gaps with his imagination, but modern-day scientists should not. This is because the intermediates are conspicuously absent from the living forms, as well as those of the fossil record. This shows Darwin's own test disproves his theory. This is negative circumstantial evidence against Darwin's theory at its best (Figure 14-1).

THE ODDS AGAINST THE THIRD LEAP OF LIFE

Paleontologists, from their study of ancient rocks, believe that prokaryotes (bacteria) were the only living inhabitants of planet Earth for about the first 2 billion years after life appeared. Then single-celled eukaryotes appeared in the fossil record and together with prokaryotes inhabited the earth for another billion years. This made a total of about three billion years of unicellularity. Then about 500 to 550 million years ago the Cambrian explosion occurred. This phenomenon was manifest by the arrival of multi-celled eukaryotes which appeared suddenly, geologically speaking, with very little evidence of ancestral past. Recall from chapter 12 that prokaryotes reproduce in some instances as often as every 20 minutes. Single-celled eukaryotes have reproductive times on the average longer than prokaryotes. Multi-celled eukaryotes on the average have longer reproductive times then single-celled eukaryotes.

According to evolutionary theory, the gradual accumulation of variations between generations leading to speciation and ultimately to new classes or even phyla on the imaginary evolutionary tree of life, depend for the most part on the random presentation of mutations to natural selection for acceptance or rejection. This is required for the theoretical evolutionary advancement to more complex life forms, which descend modified. Theoretically, from an evolutionary point of view, significant mutations leading to complexity occur during DNA replication in both prokaryotes and eukaryotes. In sexually reproducing eukaryotes, mitosis, meiosis, and fertilization, along with beneficial mutations, theoretically, are supposed to pave the way for genetic change between one generation and the next and ultimately to speciation and beyond. Obviously, the shorter the reproduction times and the larger the population the more likely for beneficial mutations to occur and the faster evolutionary advancement to more complex forms should happen. However, this is just the opposite of what paleontologists have found. Prokaryotes with the greatest population and the shortest reproduction times, took 2 billion years to theoretically evolve the single-celled eukaryotes. Single-celled eukaryotes with average reproductive times longer than prokaryotes and with smaller populations took only one billion years to evolve the multi-celled eukaryotes. According to the interpretation of the fossil record, once a few multi-celled eukaryotes supposedly evolved, this was "quickly" followed by the appearance of about 20 phyla during Cambrian times. Multi-celled eukaryotes with the longest reproductive times and the smallest populations rapidly evolved not into just new species, but also into many new phyla as well during the relatively short Cambrian explosion. These observations are the exact opposite of what scientists would expect if evolutionary theory is responsible for the progress to complexity.

For single-celled eukaryotes to rapidly evolve into multi-celled eukaryotes would require billions of beneficial mutations to be perfectly timed, placed, and sequenced in order for evolutionary processes to explain the simultaneous origin of almost all animal phyla, which appeared on planet earth during Cambrian times. This did not occur over billions and billions of years, but geologically speaking in the blink of an eye (five million years). All basic body plans and all animal phyla with possibly one exception appeared suddenly in the Cambrian fossil record without any significant genetic predecessors (Gould, Stephen J., *Scientific American*, Oct. 1994, pp. 88-89). As noted earlier, this evidence for the fast-track appearance of multi-celled eukaryotes becomes even more of an enigma to evolutionary explanation because reproduction times in this group generally are much longer than prokaryotes or single-celled eukaryotes.

One extreme example would be a sequoia redwood tree with a reproduction time greater than 100 years. Over the thousands of years that the life in one of these trees extends, it has been estimated that it will produce 60 million seeds. It also is known that the chances of any one of these seeds to geminate is very unlikely unless it lands on mineral soil. This is greatly enhanced by a forest fire that burns off the dead vegetation such as needles, leaves and dry dead branches that have fallen and collected on the forest floor. The mature redwood tree may have bark greater than onefoot thick that insulates and protects it from the ravages of fire. Therefore a 2,000-year-old tree may withstand a fire on the forest floor so that the next year's cone and seed production will drop on mineral soil, which is a requirement for germination. Several of these seeds may sprout and develop into a mature redwood tree one thousand years later. So even if a sequoia redwood tree can produce viable seeds when younger than two centuries old, it has been estimated that out of the 60 million seeds produced during the lifetime of one of these giants, only three or four sprout successfully and grow to maturity. Therefore, the reproduction time could be as long as one thousand years depending on how often a fire occurs on the forest floor. So how could this multi-celled eukaryote evolve in such a short time? Actually, it could not.

With generally longer reproduction times and much smaller populations, evolution of multi-celled eukaryotes, if true, should be slowed way down. This would be true as pointed out earlier, because most mutations in sexually reproducing eukaryotes occur during mitosis, meiosis, or fertilization as part of a generation. With longer reproduction times in multi-celled sexually reproducing eukaryotes, the number of generations per unit of time is greatly reduced and, therefore, the chance for the number of beneficial mutations needed for evolutionary theory to advance is greatly diminished. However, during the Cambrian explosion many new body plans "suddenly" and simultaneously appeared on the biological scene. This is the exact opposite of what we would expect if the Neo-Darwin synthesis is to be considered responsible for the sudden changes that occurred. Darwin predicted that further discoveries in the fossil record would confirm the slow changes in biota over the long lapses of the ages. To date, paleontologists have made no such discoveries. The intermediate life forms that Darwin predicted would be found **if his theory were true** are conspicuously absent in the fossil record. Punctuated equilibrium also fails to explain the missing intermediates. Add to the missing intermediates in the fossil record the total inability of evolution to explain the development of true sexual reproduction, which came to its acme with the appearance of multi-celled eukaryotes, and evolutionary theory is left with a mortal wound, simply waiting to die.

For the most part, Darwin studied multi-celled sexually reproducing eukaryotes. He noted very small differences between successive generations in the same species. He extrapolated in his mind that generation after generation, these slight changes would accumulate in a given species, and over time would lead to changes in a species or even cause divisions into two or more species. Each of these non-mutated differences that he saw is now understood to be the result of the triple mixing of genetic information via meiosis and sexual reproduction. This is what actually happens 100 percent of the time between one generation and the next in nearly all multi-celled sexually reproducing eukaryotes which contain no mutations. As a result, in a large population in every generation of a given species of sexually reproducing eukaryotes, hidden genetic traits located in the genomes of certain individuals will become manifest in some of their offspring and will be beneficial for their survival in a changing and possibly unfriendly environment. Because of this, a given species can present to a changing environment some individuals whose genetic makeup can survive even under many somewhat hostile situations and thus preserve the species and life itself.

Since each of the genomes of two mating partners (excluding mutations), contains all the possibilities of random change between themselves and their offspring, there is of necessity then, a limit to the amount of change available. This limitation of course was not known to Darwin and he mistakenly assumed that the slight changes he noted between one generation and the next could accumulate over thousands, or even millions of generations. He thought the accumulation of relatively small changes when preserved by natural selection, after many generations would cause big changes in multiple directions resulting in evolution of many new species. Theoretically, from his point of view, the end result of his mistaken extrapolation was supposed to cause the plethora of biota that exists today.

The amount of random change allowed by the triple mixing of sexual reproduction can help a given species to survive in many situations where a changing environment might otherwise cause them to become extinct. This is especially true when each generation produces more individuals than space or food supplies will allow to survive. In a friendly environment where plenty of space and food abound, natural selection would not have much influence on a given species in a given generation. However, the more individuals of a given species in a given generation, the greater spread will be of the possible genetic variations between individuals caused by meiosis and fertilization in sexually reproducing eukaryotes. A few of these random genetic variations would allow some individuals with the best variations to survive in a changing environment. Witness the changing beak size and shape of Darwin's finches during a recent drought on the Galapagos. Although some speciation due to the above-mentioned processes apparently has occurred, especially with geographic isolation as noted in the archipelagos of the Galapagos and Hawaii, there is no verifiable evidence that these changes have ever produced a new class or phylum. It is the variation between generations that randomly produces some individuals with changes necessary for survival. Rather than causing a species to divide into two or more species which theoretically ultimately is supposed to lead to the evolution of different classes such as fish evolving into amphibians, these random genetic changes caused by the triple mixing process of meiosis and fertilization in sexually reproducing eukaryotes, tend to the preservation of the species rather than its evolution into others.

Remember, the finches of the Galapagos Islands though differing from those of the mainland 600 miles to the east, all remained easily recognizable finches even after thousands of generations on these islands. However, in their new environment, those original colonizing birds after arriving by chance from the continent to the east were able to survive as they took up residence on these islands. This was true even though this environment was relatively hostile when compared to the one that they had left behind on the South American mainland. The reason that they could adapt to this new environment was because any slight advantage randomly acquired by the aforementioned triple mixing of genetic material, located in a given bird, would help it to survive in its new home, just as Darwin had postulated. Those finches, which by random chance contained in their genetic makeup the best combination, were able to survive in this new somewhat hostile environment. We cannot know what those slight advantages might have been back then, but they probably were similar to the changes in beak size or shape which occurred in these same finches that helped them to survive a recent drought. In just a few generations during the drought that these islands experienced, researchers noted that the changes in beak size helped these finches to survive by enabling them to feed on seeds with harder shells than what their predecessors were able to crack to obtain food. The triple mixing of sexual reproduction produced some individual birds with the proper beak changes which helped them to survive. After the drought was over the average size and shape of the beak of these birds reverted back to those before the drought.[2] Obviously these changes were not due to mutations or the changes would have been permanent. The same observation was noted with the moths that Kettlewell studied. After their environments were cleaned up, the relative numbers of peppered gray moths and peppered black moths reverted back to normal.

Rather than being a process which many evolutionary biologists insist is mainly involved in the production of a new species, or subspecies, the triple mixing process turns out to be a guardian in most cases against speciation and an explanation for stasis as noted in the fossil record as well. This is because successful non-mutated reproduction is overwhelmingly in the majority as compared with the postulated occurrence and success rate of randomly placed, extremely rare beneficial mutations needed for evolution to advance. No wonder, Niles Eldredge noted, "*Indeed, anatomical traits do shift around a bit as time goes by. But rarely do we see progressive transformation in any one direction lasting very long. What we see instead is oscillation. Variable traits usually seem to dance around an average value*" (page 69).

Now, nearly a century and a half later after the publication of *The Origen*, no step-by-step succession of intermediates has been found to exist in the fossil record, which would connect one life form with another. The platypus, the lung fish, and the archaeopteryx may be considered by some as intermediates but each of these examples could also represent a separate and distinct and/or extinct species. Darwin predicted that if his theory were true, the number of intermediates would be inconceivably great, not just a so-called missing link discovered here and there. We have seen that because the beneficial mutation rates are exceedingly rare, this would require Darwin's prediction of very slow evolutionary progress if possible, to be correct. No such series of inconceivably great numbers of intermediate life forms have been found. The mathematical calculations observed in chapters 9 and 12 show that time available for the Neo-Darwinian processes to produce the changes needed is way too short. What Niles Eldredge and others have found as they ascended the cliffs looking for fossils was evidence of new life forms showing up "with a bang," as Eldredge described in his book *Reinventing Darwin.*

To explain these discrepancies, he and Gould proposed punctuated equilibrium. The pivotal portion of this notion is that very little, if any, evolution takes place until a catastrophic event occurs, such as when an asteroid collides with Earth, which would destroy multiple ecosystems. According to this scenario, when an event of this magnitude occurs, surviving biota on the edge of the destruction evolve very fast, or as Eldredge says, "Evolution breaks loose." He says that new kinds of biota can survive easier under these circumstances because there is no competition from the old, previously established life forms. Because the new life forms evolve "very fast," according to punctuated equilibrium, they leave practically no fossil remains, and therefore no series of connecting intermediates have been found among the fossils. But the very organisms that he was talking about, have the longest reproductive times and the smallest populations and therefore cannot evolve quickly. Now you see them; now you don't.

For evolution to break loose, happen quickly, and produce new kingdoms, phyla, classes, etc. even in a few million years, would require the beneficial mutation rate to increase by many orders of magnitude. To date, no cause has been proposed or found for an increase in mutation rates. But because all mutations are rare random events, no matter the cause, there could not be a proportional increase in beneficial mutations to speed up the supposed punctuated equilibrium process or explain the sudden appearance of new biota composed of mult-icelled eukaryotes noted in the Cambrian explosion.

To admit that the intermediate fossils are lacking in the fossil record contradicts Darwin's predictions. The contradiction is further magnified by the sudden jumps noted in the fossil record that Darwin stated could not happen if his theory was true and which mutation rates prohibit. To propose that evolution can break loose and happen quickly in a relatively short time, even geologically speaking, and leave no evidence of the needed thousands of intermediate life forms, not only puts the kibosh on Darwin, but also insults our intelligence by ignoring the mathematical improbabilities imposed by rules of genetics, mutation rates, and time constraints that this notion would have us believe. Also, instead of natural selection choosing the beneficial traits being honed to perfection by competition between individuals for food, space, and survival, punctuated equilibrium relies, for the most part, on catastrophic events to cause extinctions so that new species supposedly can evolve quickly. As a result, the postulated new species can survive because there is less competition. This lack of competition causing increased fitness totally contradicts the concept of survival of the fittest and substitutes in its place survival of the most unfit.

It is difficult to understand the dichotomy that exists between the thinking of brilliant men like Niles Eldredge or the late Stephen Jay Gould, and the reality of their own observations. In his book *The Triumph of Evolution,* on Page 20 Eldredge says, "*Repeated failure to confirm predicted observations means we have to abandon an idea no matter how fondly we cherish it, or how earnestly we may wish to believe it is true*. Recall that the hallmark of all successful scientific theories requires that predictions of the results of future observations, or experiments, must be made in order to verify its truthfulness. Four predictions derived from Darwin's theory were postulated in chapter 11, all of which have failed. This is negative circumstantial evidence at its best. However, as Stephen Hawking pointed out, it only takes one. The history of science is replete with theories that have been abandoned as negative circumstantial evidence accumulated against their truthfulness. The geocentric configuration of the universe, the flat Earth, and Einstein's cosmological constant are only three examples of many that could be cited. Even though these are in the arenas of physics and astronomy, there are other debacles in the life sciences that could be mentioned. Yet, despite all the negative circumstantial evidence that exists against evolutionary theory, Darwinism is not only cherished but also extolled by the "high priests" of evolutionary biology instead of being abandoned as Niles Eldredge told us to do when repeated failure of predicted observations occur. This attitude is not unlike the so-called educated elite of Galileo's day, who, because of their cherished axioms, refused to look through his telescope to see what he saw. Or they did look, but refused to believe. The large and growing quantity of negative circumstantial evidence against evolutionary theory bespeaks of the reasons it should be abandoned, because what could not happen did not happen. It seems to be shouting: "Mr. Eldredge, tear down that wall!"[3] To paraphrase Gerald Schroeder, cherished axioms die hard even in the presence of overwhelming contradictory evidence.

Ensconced in evolution's grasp, on Earth so round and blue,
Our sun, a small but brilliant star of gold and yellow hue,
Nestled in a spiral arm as galaxies advance,
Charles Darwin's paradigm no purpose all from chance.

Paleontologist's Best Time Frame Estimates Taken from the Fossil Record (in Cosmic Time)

-Chemical Evolution- ——— Darwinian Biological Evolution ———

Theoretically, chemical evolution evolved the first living cell from non-living chemicals in less than 0.7 billion years without any help from natural selection. This is because natural selection can only make choices between live biota competing for food & space, allowing the fittest to survive & the less fit to die. Therefore, natural selection cannot participate in choices between two non-living chemicals that can't compete for food & space in the primordial soup. Chemical evolution could not begin until after Earth had cooled, sometime after it formed, 4.5 billion years ago. It ended after the first live cell arose, about 3.8 billion years ago. It began with death & ended with life.

Less than 0.7 billion years I

Like chemical evolution, biological evolution has no intelligent guidence, memory, or goal. But biological evolution has 3 advantages over chemical evolution. 1. It starts with something alive. 2. It has a choosing mechinesm – natural selection. 3. It has mutations that can change the DNA. Chemical evolution doesn't possess any of the above. Biological evolution could not begin until chemical evolution produced the first self-replicating living cell about 3.8 billion years ago. It is obvious that the ripidity of advancement in the evolution of biological complexity will be directly proportional to the number of mutations presented in a given time segment to natural selection to accept or reject. Biota with the shortest reproduction times and the largest populations will produce these conditions.

Prokaryotes, with the shortest repoduction times & largest populations of all biota, presented natural selection with the most mutations per unit of time. Therefore, single-celled eukaryotes should have evolved quickly from prokaryotes, but took 2 billion years instead.

2 billion years II

Single-celled eukaryotes with longer reproduction times and smaller populations than prokaryotes, presented natural selection with fewer mutations per unit of time and therefore, should have taken longer to evolve multi-cellulareukaryotes but took half the time instead.

1 billion years III

The Cambrian Explosion evolved 20 new animal phyla, all multi-celled eukaryotes, with longer reproduction times & smaller populations than single-celled eukaryotes. They presented natural selection with the least number of mutations per unit of time & therefore, should have taken the longest time to evolve new species, and much longer new phyla, but took the least time instead. Beginning 500 million years ago, it lasted only–5 million years. IV

The above diagram is a summary of both chemical and biological evolution derived from the fossil record. Section I describes the extreme improbability of chemical evolution, absent the selecting mechanism of natural selection, to produce the first living cell in the time available. It, like biological evolution, did not have any intelligent guidance, memory, or goal. Sections II, III, and IV describe the extreme unlikelihood of biological evolution to produce advancing complexity of biota in the time available. Theoretically, the rapidity of evolving new species on up to producing new phyla, will be dependent upon the number of mutations presented to natural selection for acceptance or rejection in any given time segment. Also, from a strictly mathematical point of view, the rapidity will be directly proportional to the size of that number. However, the fossil record displays the exact opposite. From it, the rapidity of advancement in biological complexity appears to be inversely proportional to the number of mutations occurring in a given time segment. Therefore, it seems that mathematical calculations attempting to describe the rate of biological advancement derived from neo-Darwinian synthesis theory, will yield outcomes, which are the exact opposite of predictions derived from the fossil record. This simple actuarial analysis, makes it clear that advancement in biological complexity is impossible using Darwinian theoritical principles.

Theodore Johnstome, M.D.
5-19-17

SUMMARY

1. Stasis remained for two billion years after the prokaryotes appeared until the eukaryotes supposedly evolved from prokaryotes. They existed in the unicellular form for another billion years. Three billion years of unicellularity demonstrate stasis and not inch-by-million-year inch of progress that classical evolutionary theory predicts.
2. Starting about 530 million years ago, with practically no evidence of ancestral past, in a space of about five million years, the multi-cellular eukaryotes suddenly made their appearance. This phenomenon is known as the Cambrian Explosion.
3. All the present animal phyla, with possibly one exception, appeared suddenly in the Cambrian explosion. This is a giant genetic leap of astronomical proportions.
4. It is impossible for all of the huge changes to have come about using the trial-and-error method.
5. Darwin, in *The Origin*, predicted that evolution via natural selection would produce a continuum of life forms showing only the smallest increase of change between one generation and the next. This is not what the paleontologists have found as they have studied the various fossil layers.
6. Darwin stated that **if his theory were true**, there had to be many intermediate life forms that had lived on this Earth. Therefore, it was his belief that all of life had to have descended modified from a few primordial life forms, or even one. The bottom line of his theory predicted that each species was in a state of flux with natural selection as the driving force for change.
7. Niles Eldredge and other paleontologists have found no evidence of fossils that could be categorized as a series of intermediates.
8. Instead, new fossils show up in the ascending strata "with a bang," with no evidence of an ancestral past.
9. Darwin believed that his theory would be exonerated as time went by because he thought that paleontology was in its infancy at the time he wrote *The Origin*. He believed that innumerable intermediates would be found in the fossil record.
10. Because none has been found and because stasis is the norm with a given species remaining the same for millions of years with very little change, Eldredge and Gould proposed a theory of punctuated equilibrium.
11. The upshot of this theory is that new species only evolve when an ecosystem is destroyed by a natural catastrophe such as earthquakes, avalanches, volcanoes, floods, etc. Then they say that the remaining living biota around the edge evolves quickly because there is no competition to swamp them out.
12. Genetic laws will not allow rapid speciation to occur using random mutations.
13. Punctuated equilibrium contradicts Darwin's theory in two ways. #It postulates rapid speciation which he categorically stated could not happen. #Natural selection is not the driving force of speciation but natural catastrophes instead.

ADDENDUM

Psychological Repercussions of Evolutionary Theory

EVOLUTIONARY WORLD VIEW

Now let's take the position that the theory of evolution has no flaws, and has been proven beyond any question, just like most high school biology textbooks imply. This means that from the Big Bang onward, all material objects, all composed of atoms and molecules, have been produced by pure random chance. Each atom in our entire universe is composed of nothing more then canned energy. The sun, planets, moons, asteroids, and comets have all come together by pure random means along with all the billions of stars composing our galaxy and the billions of galaxies in our ever-expanding universe.

Now if evolution really happened, then every living organism on this planet, including us humans are here by pure chance and preserved by natural selection. If evolutionary history really took place, then we ultimately are the result of random beneficial mutations that came along at just the right time and which would have caused improvements to appear all along the way in our ancestral past. Natural selection supposedly would preserve all of these improvements by its non-random means. This evolutionary interpretation implies that each person on this earth has resulted from the accumulation of ancient random molecular accidents.

Our thoughts are simply caused by chemical reactions inside and/or between the nerve cells of our brain. This also implies that our present thoughts ultimately resulted from the accumulation of many billions of previous random atomic and molecular events over which we had no control. This means we don't have free will. If one takes neo-Darwinism as the basic truth or bedrock reality, then all of what we have just discussed has to be true. Not only do humans not have free will, but no one individual's thinking can be above another's. The reason, according to evolutionary theory, since everyone's thoughts have resulted from past random events over which they had no control, they must be considered equal in quality. There can be no hierarchy of human thinking. Therefore, according to the neo-Darwinian scenario,

Albert Einstein's thinking about our universe could not be above Ted Bundy's thoughts as he was raping and killing his female victims. This is because Einstein's thinking evolved by pure chance just like Bundy's.

An article written by George Liles illustrates what has just been pointed out. The article is about what William Provine teaches at Cornell University. At the time this article was written he held joint positions in both the biology and history departments at that school. All quotes in this section are from *M.D. Magazine's* March 1994 issue (Vol. 38, No. 3, pp.59-64) except one from Nature and one from Science and Consciousness identified later. "*If you really accept evolution by natural selection, Provine says, you soon find yourself face to face with a set of implications that undermine the fundamental assumptions of western civilization.*"

- *There are no gods or purposive forces in nature.*
- *There are no inherent moral or ethical laws to guide human society.*
- *Human beings are complex machines that become ethical beings by way of heredity and environmental influences, with environment playing a larger and heredity a somewhat smaller role than is commonly supposed.*
- *There is no free will in the traditional sense of being able to make uncoerced and unpredictable choices.*
- *When we die, we die - finally and completely forever.*"

"*This, science tells us and Will Provine tells his students, is what we are and all there is.*"

Provine is not the only scientist who espouses this notion of cognitive science. Melvin Konner in the May 19, 2003 issue of Nature (pages 17-18), expresses the same idea that the mind is "a survival machine with predetermined choices." He believes that free will is an illusion. Thomas W. Clark takes a similar stance in Science and Consciousness Review, May 2002, that free will is basically a self-deception.

Later in the *MD Magazine* article Provine says, "*... Nor can we reasonably expect people to behave morally by exercising free will, because free will simply doesn't exist. Genetic and environmental factors do not merely influence our moral decisions--they determine them. Yes, people do make decisions, but they make them on the basis of deterministic influences.*"

"*William Provine's argument that free will doesn't exist grows out of a naturalistic view of the world readily accepted by many scientists, in which all events, including the neurological events we call thought, are caused by either preceding events or stochastic (random) factors, or by a combination of both. Natural laws, such as the laws of physics and chemistry, describe the impact of preceding events, and the laws of probability and chaos theory describe stochasticity. Thus both factors can be studied scientifically and neither, argues Provine, nor any mix of the two, can produce free will.*"

This way of thinking is the basis of what is called the postmodern view, which is an attack on reason and promotes unrestricted tolerance. That is, since everybody's thinking has resulted from past random events over which they had no control, there can be no way of knowing what absolute truth is. They say truth is created by each individual, not discovered as a basic foundation principle. According to this interpretation of evolutionary theory, none of us makes original choices over which we ourselves have control. Actually, neither can anyone be held accountable for his or her actions. Therefore nobody's actions are good or bad because there is no standard to judge any action to be good or bad. According to this scenario, random events that occurred without memory or goal, predetermined our choices which also determine how we live our lives.

According to this scenario, since all of us have arrived here by previous random events, and since our choices are not made by free will, then everybody's thoughts or actions must be considered equal and must not only be tolerated, but extolled. The freedom to live out one's life the way he or she seems to see fit, is really a figment of that person's imagination because each individual person's lifestyle has been preprogrammed into that person by previous random stochastic events. Therefore, each individual's actions and lifestyles must be tolerated and honored. This means that there can be no community ethic but only an individual ethic with no standard outside of the individual from which he or she can pattern his or her life. This way of thinking makes us not responsible for our actions. If this is true, then humans are simply preprogrammed biological robots.

A proof against this way of thinking exists based on modern science. The Heisenberg's Uncertainty Principle, is where the more that is known about the locality of an electron the less that is known about its velocity and vice versa. This discovery of the uncertainty principle was the beginning of quantum mechanics. The upstart of all of this is that at the basic level of physical reality, not all identical causes have the same effect. Since our thoughts are nothing more than electrons and ions moving about in our brain cells, our thoughts cannot all be the result of previous random events over which we had no control. Quantum mechanics, which cannot totally predict the outcome of cause and effect at the subatomic level, should have sounded the death knell for the postmodern view (a psychological cemetery for its burial), but it hasn't.

Our brains can produce original thoughts. We can change our minds, we do have free will, and we are responsible for our actions!

Bibliography

CHAPTER 1 BIRTH PAINS OF SCIENCE

[1] Eldredge, Niles, *The Triumph of Evolution*, 2000, W. H. Freeman and Co., New York, NY, p. 20

[2] A. *Microsoft Encarta 2006, "Aristotle."*

B. Appleyard, Understanding the Present, Doubleday, New York, New York. Originally published by Pan Am books LTD,1992, pp. 22-23.

C. Layzer, David, *Constructing the Universe,* 1984, Scientific American Books, Inc., pp.2-4.

[3] A. *World Book Encyclopedia* Vol. 15, 1968, "Ptolemy," Field Enterprises Educational Corp. p. 754.

B. *Microsoft Encarta 2006*, "Ptolemy."

[4] Appleyard, pp. 20-29.

[5] *Microsoft Encarta 2006*, "Copernicus, Nicolaus."

[6] A. *Microsoft Encarta, 2006*, "Kepler, Johannes."

B. Hummel, Charles E., *The Galileo Connection*, Inter-Varsity Press, 1986, Downers Grove, Ill., pp. 65-77.

[7] Appleyard, p. 46.

[8] *Microsoft Encarta 2006*, "Galileo."

[9] A. *Microsoft Encarta 2006, "Isaac Newton."*

B. Shamos, Morris H., Great Experiments in Physics, 1959, General Publishing Co., Ltd., Toronto, Ontario, Canada, pp. 42-48.

C. "*The Great American Bathroom Book*," eighth printing, 1995, Lan C. England, publisher, Salt Lake City, Utah, Vol. I, pp. 205-206.

D. Newton, Isaac, "To the Reverend Dr. Richard Bentley, at the Bishop of Worchester's House, in Park Street, Westminster from Cambridge, December 10, 1692" in *Theories of the Universe*, Muntiz, Miltok, Glencoe, IL, Free Press, 1957, pp. 211-212.

[10] "Kant, Immanuel, Critique of Pure Reason," in Great Books of the Western World, Vol. 42, Kant, edited by Robert Maynard Hutchins, Chicago: *Encyclopedia Britannica*, 1952, pp. 135-137.

CHAPTER 2 DARWIN

[1] Browne, Janet, (1995) *Charles Darwin Voyaging*, Princeton University Press, Princeton, NJ, pp. 6-7.

[2] Browne, p. 20.

[3] White, Michael, Gribben, John, (1997) *Darwin, A Life in Science*, Plume published by Penguin Group, New York, NY, pp. 8-9.

[4] White, Gribben, pp. 10-11.

[5] White, Gribben, p. 11.
[6] White, Gribben, p. 11.
[7] White, Gribben, p. 12.
[8] White, Gribben, p. 13.
[9] Browne, p. 35.
[10] Browne, p. 55.
[11, 12, 13] 11) Browne, p. 61. 12) White, Gribben, p. 14. 13) Browne, p. 49.
[14] White, Gribben, p. 14.
[15] Browne, p. 49.
[16] Browne, p. 62.
[17] Browne, p. 64.
[18] Browne, p. 67.
[19] Browne, p. 69.
[20] Browne, pp. 72-73.
[21, 22] 21) Browne, pp. 75, 82-83. 22) White, Gribben, pp. 17-18.
[23] Browne, pp. 88-90.
[24] Browne, pp. 90-93.
[25] Browne, p. 99.
[26] Browne, p. 103.
[27] Browne, p. 108.
[28] Browne, p. 101.
[29] Browne, pp. 109-110.
[30] Browne, p. 111.
[31] Browne, pp. 111-114.
[32] White, Gribben, p. 24.
Browne, p. 118.
[33] Browne, pp. 121-123.
White, Gribben p. 23.
[34] Browne, pp. 134-143.
[35] White, Gribben, p. 52 and Browne, p. 145.
[36] Browne, pp. 145-156.
[37, 38] 37) Browne, p. 159. 38) White, Gribben, p. 51.
[39] White, Gribben, p. 52.
[40] Darwin, Charles, *The Voyage of the Beagle*, Nal Penquin Inc., New York, NY, p. 1, Browne, p.186.
[41] White, Gribben, p. 96.
[42] White, Gribben, p. 80.
[43] Browne, pp. 66.
[44] Browne, pp.262-268 & 275-295.
[45] Browne, pp. 296-305.
[46] A. Darwin, Charles, *Origin of Species*, Chapter XI, Geographical Distribution.
B. Darwin, *The Voyage of the Beagle*, Chapter Galapagos Archipelago.
C. Browne, pp. 296-320.
[47] White, Gribben, pp. 78-80.
[48] White, Gribben, pp. 80, 99.
[49] Browne, p. 388.

CHAPTER 3 BIOLOGY 101

[1] A quote from lecture 3 on the Origins of Life by Professor Robert M. Hazen, the Teaching Company.

[2] Johnson, George B., *Biology, Visualizing Life*, Holt, Rinehart, and Winston, Publishing Inc., Austin, TX, P.179- 192.

[3] Microsoft Encarta 2006, from articles about each of the four scientists. Scientific American, April 3, 2003, Vol. 288, No 4, pp.66-69.

[4] Kleinsmith, Lewis J., Kish, Valerie M., *Principles of Cell and Molecular Biology*, 1995, 2nd ed., HarperCollins College Publishers, New York, NY, p.90.

[5] Johnson, George B., *Biology, Visualizing Life*, Holt, Rinehart, and Winston, Publishing Inc., Austin, TX, P.56.

[6] Ibid, p. 204.

[7] Kleinsmith, Kisk, p 67.

[8] Ibid, p. 81.

[9] Johnson p. 45.

[10] Bailey, Jill, *Genetics and Evolution the Molecules of Inheritance*, Oxford University Press, New York, NY, 1995, p. 21.

[11] Ibid, p. 27.

[12] Kleinsmith, Kisk, pp. 83-88.

[13] Ibid, pp. 19-27.

CHAPTER 4 BIOLOGY 102

[1] Kleinsmith, Lewis J., Kish, Valerie M., *Principles of Cell and Molecular Biology*, 1995, 2nd ed., HarperCollins College Publishers, New York, NY, p. 73.

[2] Kindrew, Sir John, Editor in Chief, *The Encyclopedia of Molecular Biology,* 1944, Alden Press limited, Oxford and Northhampton, Great Britain, p. 185

[3] Stanier, Roger Y., Ingraham, John L., Wheelis, Mark L., Painter, Page R., *The Microbial World,* 1986, 5th ed., Prentice-Hall, Eaglewood Cliffs, NJ, p. 185.

CHAPTER 5 MITOSIS

[1] Kleinsmith, Lewis J., Kish, Valerie M., *Principles of Cell and Molecular Biology*, 1995, 2nd ed., HarperCollins College Publishers, New York, NY, p. 522.

[2] Kleinsmith, Kish, p. 523.

[3] Ibid, p. 519.

[4] Stanier, Roger Y., Ingraham, John L., Wheelis, Mark L., Painter, Page R., *The Microbial World,* 1986, 5th ed., Prentice-Hall, Eaglewood Cliffs, NJ, p. 309.

[5] Kleinsmith, Kish, pp. 544-546.
Bailey, Jill, *Genetics and Evolution the Molecules of Inheritance*, Oxford University Press, New York, NY, 1995, p. 58.

[6] A very short collage of pertinent biographical information regarding several early scientist gathered from Wikipedia and Johnson, George B., *Biology, Visualizing Life*, Holt, Rinehart, and Winston, Publishing Inc., Austin, TX, P. 242.

[7] Kleinsmith, Kish, p. 523
[8] Kleinsmith, Kish, p. 522-523.
[9] Ibid, p. 73.
[10] Ibid, p. 523.
[11] Ibid, pp. 534-545.
[12] Ibid, pp. 541.
[13] Ibid, pp 543-544.
[14] Ibid, p. 544.

CHAPTER 6 MEIOSIS

[1] Kleinsmith, Lewis J., Kish, Valerie M., *Principles of Cell and Molecular Biology*, 2nd ed., HarperCollins College Publishers, New York, NY, p. 548.
[2] www.mhhe.com/biosci/cellmicro/nester/graphichs/nester3ehp/common/mcclint.html
[3] Ibid.
[4] Ibid.
[5] www.biocrs.biomed.brown.edu/Books/Essays/JumpingGenes.html
[6] www.cap.uni-muenchen.de/fgz/portals/biotech/timetable.htm
[7] www.cshl.org/public/mcclintock.html
[8] www.mhhe.com/biosci/cellmicro/nester/graphichs/nester3ehp/common/mcclint.html
[9] Kleinsmith, Kish, pp. 548-553.
[10] Hazen, Robert M., George Mason University, *The Joy of Science,* Teaching Co., Lecture 49.

CHAPTER 7 MUTATIONS

[1] EcoCye: *Encyclopedia of E. Coli Genes and Metabolism, Retrieved from: (ecocye.pangeasystems.com/ecocye/ecocyc.html).*
[2] Pennisi, Elizabeth, *Science*, "Why Do Humans Have So Few Genes?," Vol 309, No. 5731, 1 July 2005, p. 80.
[3] Bailey, Jill, *Genetics and Evolution the Molecules of Inheritance*, 1995, Oxford University Press: New York, NY, p. 35.
[4] Johnson, George B., *Biology, Visualizing Life*, 1994, Holt, Rinehart, and Winston, Austin, TX, p. 127.
[5] Bailey, p. 35.
[6] Kendrew, Sir John, Editor in Chief *The Encyclopedia of Molecular Biology*, 1994, Alden Press Limited, Oxford and Northampton, Great Britain, p. 691.
[7] Kleinsmith, Lewis J., Kish, Valerie M., *Principles of Cell and Molecular Biology*, 1995, 2nd Ed., HarperCollins College Publishers, New York, NY, pp. 90-91.
[8] Bailey, p. 96.
[9] Ibid, p. 97.
[10] Ibid, p. 35.
[11] Kleinsmith, Kish, p. 92.
[12] Ibid, pp. 95-96.
[13] www.factmonster.com/ipka/A0801697.html
[14] Ibid.
[15] Kleinsmith, Kish, p. 96.

[16] *Microsoft Encarta 2006,* Tonegawa Susumu.
[17] Bailey, p. 97.
[18] Ibid, p. 97.
[19] Kleinsmith, Kish, p. 77.
[20] Kendrew, p. 298.
[21] Ibid, p. 241.
[22] Ibid, p. 477.

CHAPTER 8 HOW MUCH TIME?

[1] A. Hawking, Stephen W., *A Brief History of Time*, 1988, Bantam Books, New York, NY, pp. 20-34.
B. Schwinger, Julian, Einstein's Legacy, the Unity of Space and Time, 1986, Scientific American Books, pp. 20, 37-39.
C. Einstein, Albert, Relativity, the Special and General Theory of, Crown Publishers, Inc., New York, 15th ed., 1952, pp. 66-70.
D. Microsoft Encarta, 2006, "Einstein."
[2] Dyson, F. W., Eddington, A. S., Davidson, C., "A Determination of the Deflection of Light by the Sun's Gravitational Field, from observations made at Total Eclipse of May 29, 1919," in Philosophical Transactions of the Royal Society of London, Series A, 220, 1920, pp. 291-333.
[3] Hawking, p. 32.
[4] Dewdney, A.K., 1997, Yes We Have No Neutrons, John Wiley and Sons NY pp.99-102.
[5] A. Hawking, p. 39-46.
B. Friedmann, Alexandre, "Uber die Krummung des Raumes," in Zeitschrift Fur Physik, 10 (1922) pp. 377-386.
C. Friedmann, Uber die Moglichkeit einer Welt mit konstanter negativer Krummung des Raumes," in Zeitschrif Fur Physik, 21, (1924), pp. 326-332.
[6] Osterbrock, Gwinn, Brashear, Scientific American, "Edwin Hubble and the Expanding Universe," July 1993, Vol. 266, pp. 84-89.
[7] Douglas, A. Vibert, "Forty Minutes with Einstein," in Journal of the Royal Astronomical Society of Canada, 50. (1956), p. 100.
[8] A. History of Women in Astronomy, The Astronomical Society of the Pacific, text by Sally Stephens (cannon. sfsu.edu/gmarcy/scwa/history/Leavitt.html).
B. Microsoft Encarta, 98, "Leavitt.
[9.] A. http://arxiv.org/abg/1212.5225 B. Planck Collaboration (2013) "Planck 2013 results. XVI. Cosmological Parameters" submitted to Astronomy and Astrophysics. C. April 30, 2013. science.nasa.gov/Planck.
[10.] Hawking, Stephen W., 1988. A Brief History of Time, p. 40
[11.] Peebles, P.J.E., (October 1994)."The Evolution of the Universe". Scientific American 271 (4) p. 54.
[12.] Halliday, David, Resnick, Robert, 1970, Fundamentals of Physics, John Wiley & Sons, Inc. New York, p. 662.
[13.] A. Hawking, et al, p. 151. B. Cosmic Horizons: Astronomy at the Cutting Edge, Edited by Steven Soter, Neil de Grasse—Tyson: a Publication of the New Press American Museum of Natural History.
[14.] Gamow, George. "Expanding the Universe and the Origin of Elements," in Physical Review, 70. (1946), pp. 572-573.
[15.] Alpher, Ralph A. and Herman, Robert C. "Evolution of the Universe," Nature, 162. (1948), pp. 774-775.
[16.] Hawking, p. 41.
[17.] Ibid, p. 41.

[18] Ibid, p. 116, and Schroeder, Gerald L. Science of God, 1997, p. 62.
[19] Ross, Hugh, The Fingerprint of God, Second Edition, Promise Publishing Co. 1991, p. 216.
[20] Peebles, P. James E., Schramm, David N., Turner, Edwin L. and Kron, Richard G., Scientific American, October 1994, Vol. 271, Number 4, p. 54.
[21] Serway, Raymond A. and Vuille, Chris, College Physics Tenth Edition, Cengage Learning, 2015, p. 340.
[22] Big Bang models back to Planck time. http://hyperphysics.phy-astr.gsu.edu/hbase/astro/planck.html. Click on the topic—Models of Earlier Events and them Quark Confinement.
[23] Fixsen, D.J. (December 2009) "The temperature of Cosmic Microwave Background," The Astrophysical Journal, 707 (2):916-920.
[24] Serway p. 1017.
[25] A. http://arxiv.org/abg/1212.5225. B Planck Collaboration (2013) "Planck 2013 results. XVI. Cosmological Parameters" submitted to astronomy and Astrophysics. C. April 30, 2013, Science.nasa.gov/Planck.

CHAPTER 9 PRIMORDIAL SOUP AND LIFE'S ORIGIN

[1] Darwin, F., *The Life and Letters of Charles Darwin*, 1888, Vol. 3, John Murray, London, p. 18.
[2] Strickberger, Monroe W., 1996. *Evolution*, Jones & Bartlett Publishers, Boston, London, Singapore, p. 115.
[3] Dawkins, Richard, 1996, *Climbing Mount Improbable*, W. W. Norton & Co., New York, NY, & London, England, p.77.
[4] Orgel, Leslie, E., *Scientific American*, "The Origin of Life on Earth," October 1994, Vol. 271, No. 4, pp. 78-79.
[5] Kauffman, Stuart, *At Home in the Universe*, 1995, Oxford University Press, New York, NY, & Oxford, England, p. 36.
[6] Orgel, Leslie, E., *Scientific American*, October 1994, p. 80.
[7] Bernstein, Max P.; Sandford, Scott A.; Allamandola, Louis J., *Scientific American*, "Life's Far-Flung Raw Materials," July 1999, p. 46.
[8] Orgel, *Scientific American,* October 1994, p. 78.
[9] Ibid, p. 81.
[10] Ibid, p. 81.
[11] Ibid, p. 80.
[12] Ibid, p. 82.
[13] Gould, Stephen J., Scientific American, Oct. 1994, pp. 86-87.
[14] Denton, Michael; Gould, Stephen J., *Evolution, a Theory in Crisis*, 1985, Adler and Adler, New York, NY, p. 250.

CHAPTER 10 THE THEORY OF BIOLOGICAL EVOLUTION

[1] White, Michael; Gribben, John; 1997, *Darwin, A Life in Science*, Penguin Group, New York, NY, p.115.
[2] Browne, Janet, 1995, Charles Darwin, Voyaging, Princeton University Press, Princeton, NJ, p. 393.
[3] White, Gribben, The Family Tree Diagrams on the pages preface to the book.
[4] Ibid.
[5] Ibid, p.103.
[6] Browne, pp. 471-480.
[7] Browne, pp. 363, 517.

[8] Darwin, Charles, 1996, *The Origin of Species,* Chapter "Geographical Distribution," Oxford University Press, Oxford, New York.
[9] Browne, pp. 386-388.
[10] Darwin, Charles; Chapter "Natural Selection."
[11] Browne, pp. 521-523.
[12] White, Gribben, 1997, pp. 96-97.
[13] Encarta, 2006, *Walter S. Sutton.*
[14] Encarta, 2006, *Thomas H. Morgan.*
[15] Encarta, 2006, *Hugo de Vries.*
[16] Bailey, Jill, *Genetics and Evolution The Molecules of Inheritance*, 1995, Oxford University Press, pp. 13-14.
[17] Dawkins, Richard, *The Blind Watchmaker*, 1996, W. W. Norton & Co., Inc., New York, NY, p. 43.
[18] Strickberer, Monroe W., *Evolution*, 1996, Jones and Bartlett Publishers International, London, England, p. 485.
[19] Johnson, George, B., 1994, Biology Visualizing Life, Holt, Rinehart and Winston Auston, p. 18.
[20] Weiner, Jonathan, *The Beak of the Finch*, 1994, Alfred A. Knopf, Inc., New York, NY pp. 121-122.

CHAPTER 11 CONFINING BOUNDARIES

[1] Cloud, P., *Oasis in Space*, 1988, W.W. Norton, New York, NY, p. 167.
[2] Allegre, Clause S.; Schneider, Stephan H., *Scientific American*, Evolution of the Earth, Oct. 1994, p. 69.
[3] Professor Robert M. Hazen, Lecture 20, "The Joy of Science," The Teaching Company, 2001.
[4] Serway, Raymond A.; Faughn, Jerry S., *College Physics,* 3rd ed. 1992. Sanders College Publishing, Fort Worth, TX., p. 965.
[5] White, Michael; Gribbin, John, *Darwin, A Life in Science*, Plume, Published by the Penguin Group, April 1977, p. 80.
[6] Kettlewell, H. B. D., *Scientific American*, "Darwin's Missing Evidence," 1959. vol. 201, number 3 p. 48-53.
[7] Behe, Michael, J., *Darwin's Black Box*, 1996, Free Press, New York and London, pp. 77-97.
[8] Browne, Janet, *Charles Darwin Voyaging*, 1995, Princeton University Press, Princeton, New Jersey, pp. 471-480.
[9] Browne, Janet, op crt. p. 471.
[10] Hazen, Robert, Lecture 49.

CHAPTER 12.THE APPEARANCE OF SINGLE-CELLED EUKARYOTES

[1] Kleinsmith, Lewis J., Kish, Valerie M., *Principles of Cell and Molecular Biology*, 1995, 2nd Ed., HarperCollins College Publishers, New York, NY, p. 90.

CHAPTER 13 FOSSILS

[1] Rothery, David A., *Teach Yourself Geology*, 1997, NTC Publishing Group, Lincolnwood, IL, pp. 2, 72.
[2] *Dictionary of Geology*, 1984, Doubleday, 3rd Ed. Prepared by the American Geological Institute, New York, N.Y., Robert L. Bates, and Julia A. Jackson, Editors, pp. 136, 219, 289, 311, 471.

[3] Rothery, Chapters 8 & 9.

[4] Rothery, p. 107.

[5] *Dictionary of Geology*, p. 172, 295.

[6] Rothery, pp. 144-145.

[7] Gilluly, James; Waters, Aaron C.; Woodford, A. O., *Principles of Geology*, 1968, 3rd Ed., W. H. Freeman and Co., San Francisco, CA, & London, England, p. 35.

[8] Rothery, Chapters 3, 4, & 5.

[9] Morton, Glenn, *The Geological Column*, February 17, 2001, retrieved from www.museum.state.il.us/isas/kingdom/geo1002.html

[10] *Dictionary of Geology*, p. 546.

[11] *Dictionary of Geology*, p. 77.

[12] Gould, Stephen Jay, 1989, *Wonderful Life*, W. W. Norton & Company, New York, NY, pp. 61-62.

CHAPTER 14 CONFOUNDING BOUNDARIES

[1] Darwin, C., (1858) in a letter to Asa Gray, 5 September, 1857, Zoologist 16: 6297-99. See page 6299.

[2] Gibbs, H. Lisle, Grant, Peter R., *Nature,* Vol 327, 11 June 1987, pp. 511-513.

[3] A paraphrase taken from a speech of President Ronald Regan made while standing before the Berlin Wall.